MW00680498

AS9101D Auditing for Process Performance

Also available from ASQ Quality Press:

A Practical Field Guide for AS9100C
Erik Valdemar Myhrberg, Dawn Holly Crabtree, and Rudy "RE" Hacker

The ASQ Auditing Handbook, Third Edition
J.P. Russell, editing director

Quality Audits for Improved Performance, Third Edition
Dennis R. Arter

The Making of a World-Class Organization
E. David Spong and Debbie J. Collard

ISO 9001:2008 Explained, Third Edition
Charles A. Cianfrani, John E. "Jack" West, and Joseph J. Tsiakals

ISO Lesson Guide 2008: Pocket Guide to ISO 9001:2008, Third Edition
J.P. Russell and Dennis R. Arter

ISO 9001:2008 Internal Audits Made Easy: Tools, Techniques, and Step-by-Step Guidelines for Successful Internal Audits, Second Edition
Ann W. Phillips

Process Driven Comprehensive Auditing: A New Way to Conduct ISO 9001:2008 Internal Audits, Second Edition
Paul C. Palmes

ISO 9001:2008 Interpretive Guide for the Design and Construction Project Team (e-Book)
Prepared by members of the ASQ Design and Construction Division and edited by John R. Broomfield

The Quality Toolbox, Second Edition
Nancy R. Tague

Mapping Work Processes, Second Edition
Bjørn Andersen, Tom Fagerhaug, Bjørnar Henriksen, and Lars E. Onsøyen

Root Cause Analysis: Simplified Tools and Techniques, Second Edition
Bjørn Andersen and Tom Fagerhaug

The Certified Manager of Quality/Organizational Excellence Handbook, Third Edition
Russell T. Westcott, editor

To request a complimentary catalog of ASQ Quality Press publications, call 800-248-1946, or visit our website at http://www.asq.org/quality-press.

AS9101D Auditing for Process Performance

Combining Conformance and Effectiveness to Achieve Customer Satisfaction

Chad Kymal

ASQ Quality Press
Milwaukee, Wisconsin

American Society for Quality, Quality Press, Milwaukee 53203

Printed in the United States of America
17 16 15 14 13 12 11 5 4 3 2 1

Library of Congress Cataloging-in-Publication Data

Kymal, Chad.
AS9101D auditing for process performance : combining conformance and effectiveness to achieve customer satisfaction / Chad Kymal.
p. cm.
Includes bibliographical references and index.
ISBN 978-0-87389-807-2 (hardcover : alk. paper)
1. Aerospace industries—Standards. 2. Aeronautics—Specifications. I. Title.

TL671.1.K96 2011
657'.867—dc22 2011013141

ISBN: 978-0-87389-807-2

Publisher: William A. Tony
Acquisitions Editor: Matt T. Meinholz
Project Editor: Paul O'Mara
Production Administrator: Randall Benson

ASQ Mission: The American Society for Quality advances individual, organizational, and community excellence worldwide through learning, quality improvement, and knowledge exchange.

Printed on acid-free paper

Quality Press
600 N. Plankinton Ave.
Milwaukee, WI 53203-2914
E-mail: authors@asq.org

The Global Voice of Quality™

Table of Contents

List of Figures and Tables

Preface

The aerospace industry—unlike some other industries who have been badly hurt in the economic downturn—has not only survived, but all indications point toward a period of strong growth. Most organizations in the aerospace industry are already looking ahead to tremendous opportunities. The challenges are formidable, but the benefits are huge.

In the next two decades, the aerospace industry will undergo dramatic changes for three reasons: the need for improvement, marketplace growth, and awareness of major technological changes for energy efficiency in the aerospace and defense industries. The aerospace industry has few competitors, not only at the OEM (original equipment manufacturer) level, but also in aerospace subsystems such as engines, hydraulics, and/or wing composites. This lack of competition has resulted in an average ppm (parts per million) level of 30,000 ppm and an on-time delivery rate of 50% or less. The need for improvement will be precipitated by new competitors from organizations traditionally associated with the automotive industry in the United States seeking to diversify their operations at the tier I level, as well as competitors from India and China looking to enter the lucrative and high-growth aerospace industry at both the OEM and tier I and II levels.

The aerospace industry will be fueled by the economies of China and India, which are expected to emerge as the largest market, followed by North America and then Europe (see table). The need for localization of aerospace content and adherence to the governmental policies of China and India will lead to aerospace joint ventures (JVs) and industries that will steadily transform into competitors of the future.

Along with this expected market growth and a level of performance that would be considered less than world-class, there is huge rush of new technologies in the aerospace industry, including electronics, composites, and other lightweight technologies. The opportunities and growth in the

Aerospace industry demand by region—2010 to 2029.		
Region	**New airplanes**	**Value ($B)**
Asia Pacific	10,320	1,320
North America	7,200	700
Europe	7,190	800
Middle East	2,340	390
Latin America	2,180	210
CIS*	960	90
Africa	710	80
Total	**30,900**	**3,590**

*Commonwealth of Independent States.

market, coupled with inefficient companies, new technologies, and new or inexperienced competitors, bring large risks but great rewards.

Risks will be prevalent for those entrenched players who either are resistant to change or are not able to quickly adapt to the new realities of the marketplace. Opportunities abound for those newer and more nimble competitors who are able to embrace change in both technology and the supply chain.

This is the proper prism through which to view AS9100 Revision C. Many organizations in the industry have embraced these changes, and, to a large extent, the need for change is reflected in the AS9100C standard through the introduction of the concept of risk management. Process auditing and the idea of prioritizing the audit are key themes emerging out of the new AS9100 auditing standard. In fact, at Omnex we are bold enough to say that an aerospace quality management system (AS9100) implemented and audited using a customer and process focus will help organizations transform to a large degree. However, the focus of the implementation will need to change from "conformance" to "best-in-class."

Key changes to the AS9100 auditing process include the following:

- Introduction of the PEAR (Process Effectiveness Assessment Report) tool
- Process orientation of the organization
- Customer metrics and measurements

- Project management with risk assessment built throughout the company (not only in design)
- Linkages of characteristics (special characteristics, critical characteristics, and key characteristics, including the flow-down of metrics)
- Importance of customer requirements (customer requirements via standards and contract)
- Configuration management

As author of this book, I aim to accomplish two overall goals: first, to help aerospace organizations improve through effective audit practices, and second, to help auditors improve their overall auditing skills.

NOTE TO THE READER

This book is focused on internal audits and internal quality system audits. As such, keep in mind the following:

- There are three common types of audits: system, process, and product audits. The focus of this book is on system audits. Product and process audits are important and will be the focus of future books and/or articles. At Omnex, we couple process audits with the manufacturing process flow, PFMEA, control plan, and work instructions of the process. In the coming years, we expect more and more aerospace organizations to adopt these risk management tools, making the process audit that much more important.
- Internal audits of the quality system need to be performed as stage 1 and stage 2 audits. The stage 1 audit is essential for planning and focusing the audit first on customer performance issues and then on process "performance." The stage 2 audit has many purposes, including auditing leadership, customer satisfaction, effectiveness of the system, customer performance issues, and other important topics. Read this book with the idea that the internal audit process is focused on improving your overall organization. The outcomes of AS9100C and AS9101D process-focused audits are nonconformities that address fundamental issues within the organization. Therefore, nonconformities such as document control issues, document updates, and people "not trained" are not the outcomes expected for a properly executed audit as discussed in this book.

- Chapters 7 and 8 discuss the stage 1 and stage 2 audits. Both of these chapters are supported by audit checklists for the internal audit. The audit checklists in Appendixes A and B follow the same steps described for the audit process in Chapters 7 and 8.
- Appendix C (on CD-ROM included with the book) includes an audit checklist that encompasses the Objective Evidence Report (OER, AS9101D). Though the checklist is clause-based, the audit should be process-based. Auditors auditing processes should cross-reference the clauses in the audit plan. *Note:* Internal auditors require guidance for conducting internal audits, and additional guidance is provided through the audit checklist in Appendix C.
- Internal quality system audits that drive performance and organizational improvement require the support of management for effective implementation. This support is key as the AS9101D audit for process performance is conducted in your organization.

Overall, this book is designed to be read progressively. The key chapters include Chapter 6 (Understanding Audit Trails), Chapter 7 (Stage 1 Audit), and Chapter 8 (Stage 2: On-Site Audit).

Chad Kymal
CTO and Founder, Omnex Inc.

1

History of Aerospace Industry Standards

The aerospace industry standards as we know them today had their start in the mid-1990s when a team from AlliedSignal, Pratt & Whitney, Boeing, Rockwell Collins, GE Engines, Rolls-Royce Allison, Lockheed Martin, Sundstrand, McDonnell Douglas, and Northrop Grumman came together to create an aerospace industry standard under the auspices of the American Society for Quality (ASQ) and Society for Automotive Engineers (SAE). The resulting standard was the AS9000:1997 standard.

AS9000:1997

The 1997 version of the standard was designed using various standards from the participating companies, including D1-9000, ISO 9001, DOD, FAA, NASA, MIL-Q, MIL-1, and other standards prevalent in the aerospace and defense industries. The endeavor was primarily a U.S. effort resulting in the first AS9000 standard, which was released in 1997.

Almost immediately, an effort was made to create an international aerospace standard under the oversight of ISO Technical Committee 20—Aircraft and Space Vehicles. The United States convened the committee, which included members from Brazil, China, France, Germany, Japan, Mexico, the United Kingdom, and the United States. Three separate standards were merged to form the first version of the AS9100 standard structured around the ISO 9001 standard. This standard became the first aerospace world quality standard, AS9100:1999.

AS9100:1999 (REVISION A)

This is the first aerospace standard that was based on and built around ISO 9001, the international standard for quality management systems requirements. At the time of publication, AS9100:1999 added 83 additional requirements over and above ISO 9001, including 11 amplifications of some areas of the standard. The greatest emphasis was placed on design control, process control, purchasing, inspection and control, and control of nonconformances. The rationale of the writing committee was that ISO 9001 did not go far enough to capture regulatory requirements or the importance of safety, reliability, or maintainability. Second, it was necessary to capture aerospace supplemental requirements agreed to at an international level. This first standard was the predecessor to the current AS9100C standard, for all practical purposes.

At the same time as AS9100:1999 was released, the International Aerospace Quality Group (IAQG) was formed (see Figure 1.1). All responsibilities were transferred from the WG11, who authored the AS9100:1999 Revision A standard, to the newly formed IAQG. AS9100:1999 was published in the United States, Europe, and Japan with three different standard numbers—a practice that still prevails to this day.

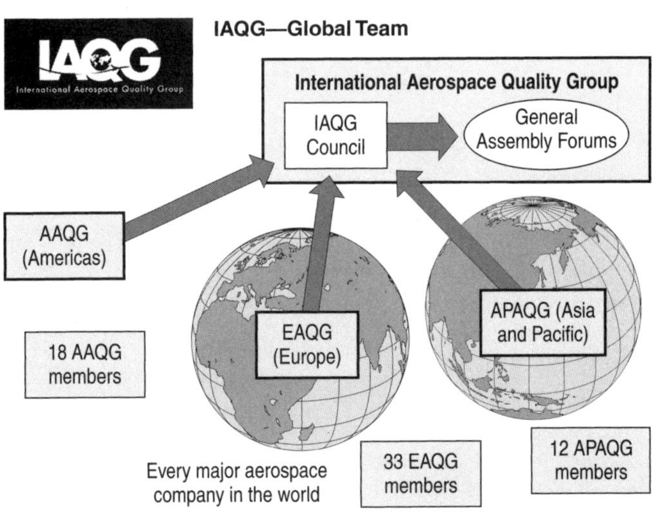

Figure 1.1 International Aerospace Quality Group (IAQG) global organization.

Source: IAQG website *AS9100 Changes,* page 3.

AS9100 REVISION B

Almost immediately, the IAQG started to work on revising AS9100 to be harmonized with the ISO 9001:2000 standard. ISO 9001:2000 was a major departure and a significant change from the older ISO 9001:1994 standard on which the AS9100:1999 standard was built. Since many of these standards were released in such tight time frames, AS9100 Rev A consisted of two separate sections so it could be integrated with either ISO 9001:2000 or ISO 9001:1994. This allowed organizations to work with the AS9100 standard regardless of their current ISO 9001 registration. It wasn't until 2003 that the IAQG released AS9100 Rev B, which simply removed the section on ISO 9001:1994 requirements.

AS9101—AUDIT CHECKLIST

In conjunction with AS9100 Rev B, the AS9101 audit checklist was released. This marked the inception of a common audit checklist for the aerospace industry. It was written to cover all requirements of both AS9100 and ISO 9001. Certification bodies were required to use this checklist when auditing to AS9100. The release of the checklist also coincided with the formation of a certification scheme developed by geographical area. The idea was to create a system that granted the ability to share audit results, hence reducing the multiple audits prevalent in the aerospace industry. The goal of the audit system was to conduct one audit that would provide confidence to the industry that multiple additional audits are not necessary. The OASIS database was also launched at this time.

AS9104

In order to standardize all requirements for auditors, registrars, and accreditation bodies, the IAQG developed a new standard titled AS9104. This standard was released in June of 2001 and included the following important considerations:

- Requirements for accreditation bodies (ABs) and certification bodies (CBs)
- Requirements for auditors
- Training requirements explicitly stated

- Requirements for reporting results of audits
- Minimum audit times and guidelines
- Industry involvement
- Requirement of suppliers to notify OEMs of status of registration and any changes thereto
- Requiring that problems must be reported to registrars
- Tracking suppliers versus registrars versus performance

FOUNDATION YEARS 2000 AND 2001

2000 and 2001 were important years for the AS9100 series of standards. In a short period of time the building blocks of the present-day standards were put in place, as was the IAQG and its three regional groups—the Americas Aerospace Quality Group (AAQG), the European Aerospace Quality Group (EAQG), and the Japan Aerospace Quality Group (JAQG). Many of the standards as we know them today had their start in 2001, including the AS9100 standard integrated with ISO 9001, the AS9101 audit checklist, and the first attempt to standardize requirements for auditors, registrars, and accreditation bodies with the AS9104 standard. Other related standards, such as *AS9102—Aerospace First Article Inspection Requirements* and *AS9103—Variation Management of Key Characteristics,* had already been in use by some companies in the aerospace community.

It was also at this time that a version of AS9100 for distributors was published in Europe as EN9120. EN9120 was accompanied by its own checklist, EN9121. At that time, the United States had a few competing distributor standards, including ASA-100 and AS7103. The standard for aviation maintenance organizations, AS9110, was not even being considered at this time. The foundation for the creation of the AS9100 standard can be seen in Figure 1.2.

AS9110 Requirements for Aviation Maintenance Organizations and *AS9120 Quality Management Systems—Requirements for Aviation, Space and Defense Distributors,* the most recent additions to the AS family, were published in January 2003 and November 2002 respectively. These two standards were based on ISO 9001:2000 and focused on the specific quality requirements of the maintenance and stockist distributor segments of the aerospace industry.

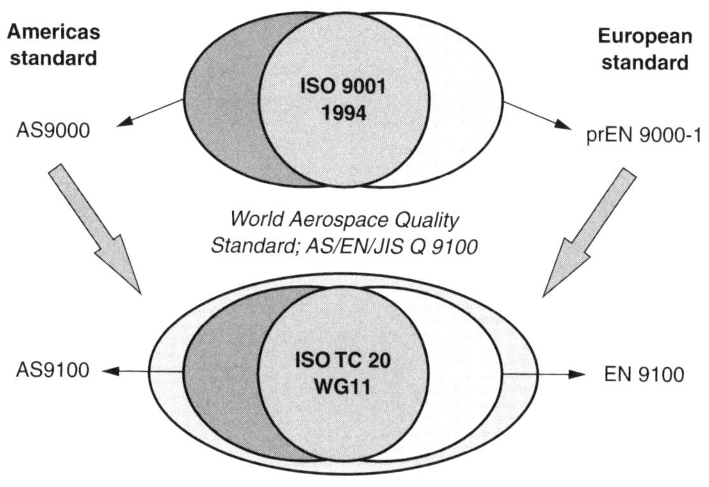

Figure 1.2 Foundation of the AS9100:1999 (revised) standard.
Source: The Koyoto AS9100 IAQG Meeting, presentation by Dale K. Gordon, March 14, 2002.

AS9100 REVISION C

The latest revision of AS9100 (Rev C) was released in January, 2009, approximately three months after the release of the ISO 9001:2008 standard. By the time AS9100C was released, IAQG had become a well-established aerospace industry group with wide acceptance in the international aerospace community. Most of the aerospace OEMs already required conformity to AS9100 for their prime contractors. Many tier 2 and tier 3 aerospace organizations had also adapted the standard to their own use.

Beyond the incorporation of ISO 9001:2008, the biggest change to AS9100C was the expansion of scope to include land- and sea-based systems for defense applications, as well as risk management, project management, configuration management and critical items, and special requirements. The argument can easily be made that the 2009 changes (Rev C)—along with the new AS9101 auditing standard (AS9101D)—are the biggest changes to the AS9100 series of standards in a long time.

The changes to AS9100 come from ISO 9001:2008 sections on risk management and critical/special requirements, and the AS9101 auditing standard. At the time Rev B was released in 2001, the impact of the process approach versus the procedural approach of ISO 9001:2000 and

ISO 9001:1994 was minimized for many reasons, due at least in part to the quick succession of the 1999 and 2001 versions. In AS9100C, the process approach and its effect on aerospace auditing as a whole is going to come to a head. The need for risk management in the aerospace industry, along with the process approach and the audit approach of AS9101, will be among the biggest changes to tackle from both an implementation and an auditing perspective. These and other changes will be covered in detail in the next chapter.

AS9100, AS9110, AND AS9120 STANDARDS

Aerospace, Defense, and Space Industries

The AS9100C standard was written with the assistance of 18 members representing the Americas, Europe, and Asia-Pacific IAQG sectors. The objectives of the revision included incorporating all ISO 9001:2008 changes, expanding the scope to include land- and sea-based systems for defense applications, ensuring alignment with the IAQG strategy of on-time and on-quality performance, and adopting new requirements based on stakeholder needs, as well as improving existing requirements where stakeholders identified a need for clarification, including instances where a documented procedure is needed.

The scope of the AS9100C standard is aerospace, defense, and space industries. With this revision, defense industries were added to the scope of the AS9100 standard. This revision includes six additions, eight revisions and relocations, and three deletions to AS9100, on top of the changes to ISO 9001:2008. This will be covered in the next chapter.

Maintenance Organizations

The AS9110 standard was written with the assistance of 12 members from the Americas and Europe IAQG sectors representing five different countries, eight different IAQG member companies, and three different CBs. The objectives of the 9110 revision included incorporating ISO 9001:2008 changes and 9100C changes as applicable to maintenance activities, providing clarity in order to resolve interpretation issues, addressing stakeholder needs, promoting an industry standard and ensuring that this standard is compatible for use by all stakeholders, and preparing for the forthcoming safety management system implementation requirements in accordance with International Civil Aviation Organization (ICAO) directives.

The intended application of AS9110 includes:

> . . . use by maintenance organizations whose primary business is providing maintenance, repair, and overhaul services for aviation commercial and military products; and for Original Equipment Manufacturer (OEM) organizations with maintenance, repair, and overhaul operated autonomously or that are substantially different from their manufacturing/production operations.
>
> [It] is tailored for organizations with National Airworthiness Authority (NAA) repair station certification and those that provide maintenance, repair, and overhaul services for military aviation products; but the standard could significantly benefit non-certificated maintenance organizations that choose to adopt it. *(AS9110A, 1.2—Application)*

AS9110 includes nine additions and one revision. This book will cover the AS9110 changes in Chapter 3. See Table 1.1 for a comparison between AS9100, AS9110, and AS9120.

Distributors to Aviation, Space, and Defense

AS9120 was written with eight members representing the Americas, Europe, and Asia-Pacific IAQG sectors including distributors, OEMs, and associations from both production and aftermarket. The objectives of the 9120 revision included incorporating ISO 9001:2008 changes, alignment with the IAQG strategy (on-time, on-quality performance), alignment with AS9100C changes, and adopting new requirements based on stakeholder needs, as well as improving existing requirements where stakeholders identified a need for clarification, including instances where a documented procedure is needed.

The scope of AS9120 applies to distributors to the aviation, space, and defense industries. Similarly to AS9100, defense industries were added to the scope of AS9120. The standard includes six additions, five revisions/relocations, and two deletions. This book will cover the AS9120 changes in Chapter 3. See Table 1.1 for a comparison between AS9100, AS9110, and AS9120.

Common Aerospace Auditing Standard

AS9101, which was released in March, 2010, was written by 13 members, representing the Americas, Europe, and Asia-Pacific IAQG sectors, from six different countries, including eight IAQG member companies and four certification bodies.

The objectives of the 9101D revision included alignment with the IAQG strategy (on-time, on-quality performance, improved control of other party

Table 1.1 Comparison of ISO 9001:2008 to AS9100, AS9110, and AS9120.

	ISO 9001:2008		AS9100 compared to ISO 9001:2008		AS9110 compared to AS9100		AS9120 compared to AS9100
1.0	Scope	1.0	Scope General Reference that the standard includes ISO 9001:2008 requirements	1.0 1.1	Scope General References baseline maintenance requirements	1.0	Scope Same
				1.2	Application References baseline maintenance requirements		
2.0	Normative references	2.0	Normative References—Same	2.0	Normative references—Same	2.0	Normative references—Same
3.0	Terms and definitions	3.0	Terms and definitions—Additional definitions	3.0	Terms and definitions—Additional definitions	3.0	Terms and definitions—Additional definitions
4.0	Quality management system	4.0	Quality management system—References to statutory and regulatory requirements	4.0	Quality management system—References baseline maintenance requirements	4.0	Quality management system
		4.2 4.2.1	Documentation requirements General—Organization's personnel aware and have access to QM documentation and changes	4.2.2	Quality manual—Additional requirements specific to maintenance personnel under "d"		

Continued

Table 1.1 *Continued.*

	ISO 9001:2008		AS9100 compared to ISO 9001:2008		AS9110 compared to AS9100		AS9120 compared to AS9100
		4.2.4	Control of records—Added reference to suppliers			4.2.4	Control of records—Added additional records
5.0	Management responsibility	5.0	Management responsibility	5.0	Management responsibility	5.0	Management responsibility—Same
		5.2	Customer focus—Added reference to on-time delivery	5.1	Management commitment—Adds safety and safety objectives		
		5.5.2	Management representative—Added reference to unrestricted freedom to top management to resolve QM issues	5.4.3	Safety objectives—Added		
				5.5.1.1	Accountable executive manager—Added		
				5.5.1.2	Maintenance managers—Added		
				5.6.2	Review input—Added h, I, j		
				5.7	Safety policy—Added		

Continued

Table 1.1 *Continued.*

	ISO 9001:2008		AS9100 compared to ISO 9001:2008		AS9110 compared to AS9100		AS9120 compared to AS9100
6.0	Resource management	6.0	Resource management—Same	6.0	Resource management—References baseline maintenance requirements	6.0	Resource management—Same
				6.2	Human resources		
				6.2.2	Competence training and awareness—Added f, g, h		
				6.3	Infrastructure—Added d		
7.0	Product realization	7 7.1	Product realization Planning of product realization—Expanded a. quality objectives	7.0	Product realization—Minor additions related to special product requirements	7.0	Product realization
		7.1.2	Risk management—Added				
		7.1.3	Configuration management—Added			7.5.2	Validation of processes for production and service provision—Delete
		7.1.4	Control of work transfers—Added			7.1.1	Design and development—Added
		7.1.1	Project management—Added			7.1.2	Risk management—Deleted

Continued

Table 1.1 *Continued.*

	ISO 9001:2008		AS9100 compared to ISO 9001:2008		AS9110 compared to AS9100		AS9120 compared to AS9100
		7.2.2	Review of requirements related to product—Added d and e			7.2.2	Review of requirements related to product—Deleted c. ability to meet defined requirements
		7.3.1	Design and development planning—Added references to distinct actvities, safety and functional objectives, and ability to produce			7.3	Design and development—Deleted
		7.3.3	Design and development inputs—Added e				
		7.3.3	Design and development review—Added c				
		7.3.6.1	Design and development verification and validation testing—Added				
		7.3.6.2	Design and development verification and validation documentation—Added				

Continued

Table 1.1 *Continued.*

	ISO 9001:2008		AS9100 compared to ISO 9001:2008		AS9110 compared to AS9100		AS9120 compared to AS9100
		7.4 7.4.1	Purchasing Purchasing process—Major revisions			7.4.1	Purchasing process—Added g. Implement controls to prevent the purchase of counterfeit and suspected unapproved parts
		7.4.2	Purchasing information—Major revisions				
		7.4.3	Verification of purchased product—Major revisions				
		7.5	Product and service provision—Major revisions				
		7.5.1.1	Product and process verification—Added				
		7.5.1.2	Control of production process changes—Added				
		7.5.1.3	Control of production equipment, tools, and software programs—Added				
		7.5.1.4	Post delivery support—Added				

Continued

Table 1.1 *Continued.*

	ISO 9001:2008		AS9100 compared to ISO 9001:2008		AS9110 compared to AS9100		AS9120 compared to AS9100
		7.5.5	Preservation of product—Major Changes				
		7.6	Control of monitoring and measuring equipment—Notes added				
8.0	Measurement analysis and improvement	8.0	Measurement analysis and improvement	8.0	Measurement analysis and improvement	8.0	Measurement analysis and improvement
		8.1	General—Added notes				Deleted section after 8.1c
		8.2.1	Customer satisfaction—Added note				
		8.2.2	Internal audit—Added note				
		8.2.3	Monitoring and measurement of processes—Added a, b, c, d				
		8.2.4	Monitoring and measurement of product—Major additions				

Continued

Table 1.1 *Continued.*

	ISO 9001:2008		AS9100 compared to ISO 9001:2008		AS9110 compared to AS9100		AS9120 compared to AS9100
						8.2.5	Evidence of conformity—Added
		8.3	Control of nonconforming product—Major additions			8.3	Control of nonconforming product—Distributer has no authority to rework or repair
		8.5.1	Continual improvement—Note added				
		8.5.2	Corrective action—Added g, h, i				
		8.5.3	Preventive action—Added note				

certification), incorporating the AS9100C changes and aligning with ISO 17021 content, supporting the process approach for quality management systems as described in the 9100-series standards, incorporating the ICOP (Industry Controlled Other Party) resolutions and improving the value of third-party auditing, defining the requirements for all activities of the audit program, and providing an adequate audit report to stakeholders on the performance of organizational processes.

AS9101 defines the common auditing requirements for the AS9100, AS9110, and AS9120 standards. The AS9101 audit standard is a complete rewrite of the previous version it replaces, making AS9101 a process-focused and performance-based standard. In many ways, Figure 1.3 represents a summary of the changes the standard represents. The AS9101 standard and its changes will be covered in detail in Chapter 3.

SUMMARY

The aerospace standards had their start in 1997 with the advent of the AS9000:1997 standard. This represented the first effort at standardization by the aerospace industry. The next big step toward standardization was the foundation of the IAQG and the release of AS9100 Rev B.

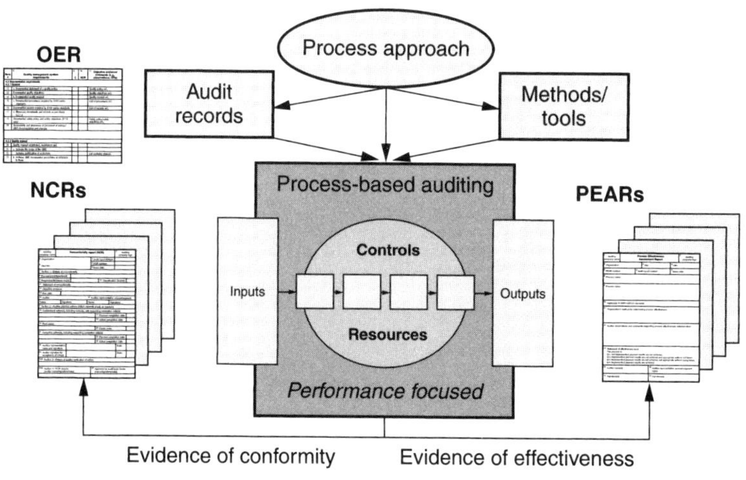

Figure 1.3 Model for process-based auditing.

Source: AS9101 model for process-based auditing, IAQG 9101:2009 change overview presentation, April 2010.

With the release of AS9100 Rev C and the common AS9101D audit standard, the push to unify the aerospace industry under a common set of standards is hitting its stride. The AS9100 series of standards and the IAQG have reached their maturity, and their influence and impact will be felt by the industry for the next 10 years. The advent of AS9100C will push customer-focused performance of "on-time and quality" and overall customer satisfaction. There will be another improvement surge in third-party auditing as well. All these changes bode well for the aerospace industry.

2

AS9100C Key Changes and the Impact to the Quality Management Systems

An AS9100-compliant quality management system (QMS) is impacted not only by the AS9100 standard, but also by the AS9101 standard. Though the AS9101D requirements are mandatory for third-party auditors and registrars, the expectations of the auditor will dictate the requirements that the organization's QMS needs to meet. This chapter will cover the key changes to the QMS and will also list a number of QMS requirements that are influenced by the AS9101D auditing standard. These and other AS9101D changes will be covered in greater detail in the next chapter.

Overall, three groups of changes impact an AS9100 quality management system: AS9100C changes, AS9101D changes, and ISO 9001:2008 changes. All three of these need to be considered by organizations implementing and auditing to AS9100C.

AS9100 CHANGES

The IAQG identifies six additions, eight revisions/relocations, and three deletions to AS9100 on top of the changes to ISO 9001:2008. The six additions they refer to are:

1. Risk (3.1): undesirable situation or circumstance
2. Special Requirements (3.2): requirements that have high risk
3. Critical Items (3.3): items (such as functions, parts, software, characteristics, processes, and so on) that have a significant effect on product realization and use of the product (including safety, performance, fit, function, producibility, service life, and so on) that require specific actions to ensure they are adequately managed

4. Customer Focus and Customer Satisfaction (5.2 and 8.2.1): focus on product conformity and on-time delivery performance
5. Project Management (7.1.1): planning and managing product realization (new product development) with acceptable risk
6. Risk Management (7.1.2): process for the management of risk in product realization

Risk, special requirements, critical items, project management, and risk management are all linked together. They are a part of the overall risk management process in an aerospace, space, or defense organization. Figure 2.1 demonstrates the interrelationship of these new additions.

Impact of Key Characteristics and Risk Management Process on the QMS

Auditors should expect a comprehensive risk management process implemented at the organizations they audit. The management of risk starts at sales for suppliers (7.2 Customer-Related Processes) or in the planning phase (7.1 Planning of Product Realization) for OEMs. See Figure 2.2 for the product realization process.

The risk management process needs to be embedded in sales or contract review, planning, product design, manufacturing and delivery, and post-delivery processes. It should include the definition of all special requirements, critical items, and key characteristics. Though key characteristics are not new to AS9100C, they act as an important designation for controlling variation on the shop floor. When auditing an organization, the auditor should look for evidence of a risk management process embedded within the product realization process that includes special requirements, critical items, and key characteristics. If the risk management process is implemented as a stand-alone process, the linkages between the risk management process and planning, sales, design, manufacturing, and first article inspection (FAI) processes must be clearly defined.

During the IAQG-sanctioned auditor transition training, examples of risk management were demonstrated through the use of design failure mode and effects analysis (DFMEA), process failure mode and effects analysis (PFMEA), control plans, and statistical studies both during FAI and after FAI for process control. Many aerospace organizations are embracing the FMEA methodology along with a structured new product development phase gate approach for the launch of new products. In fact, Omnex is currently working with some large aerospace manufacturers and their suppliers—including Bombardier and Pratt & Whitney—to introduce these tools into their new-product launch processes.

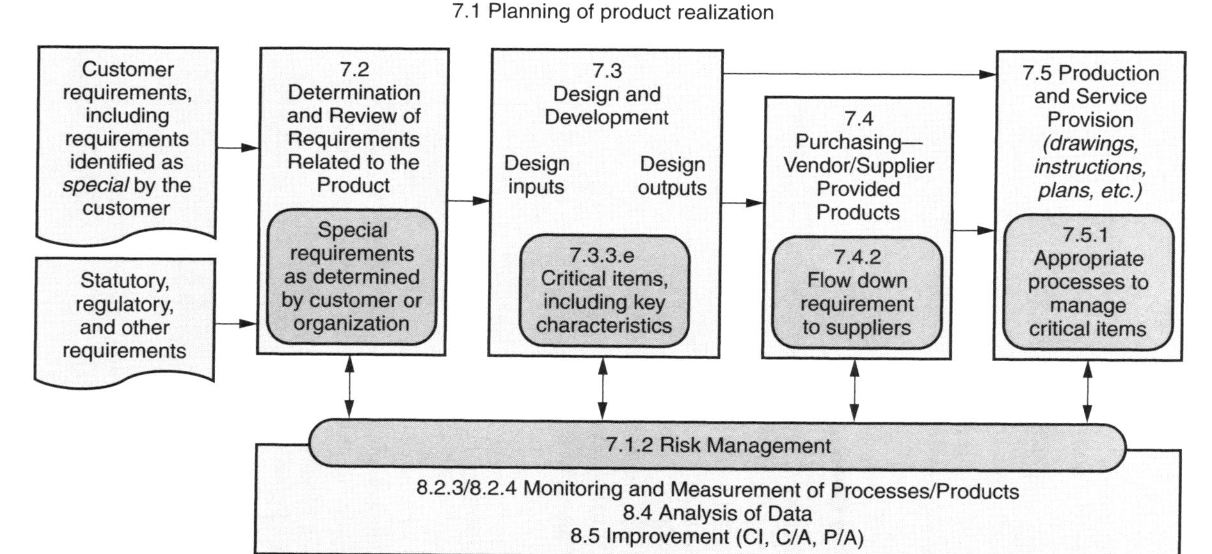

Figure 2.1 Interrelationship between special requirements, critical items, key characteristics, and risk management process.
Source: IAQG AS9100 deployment materials.

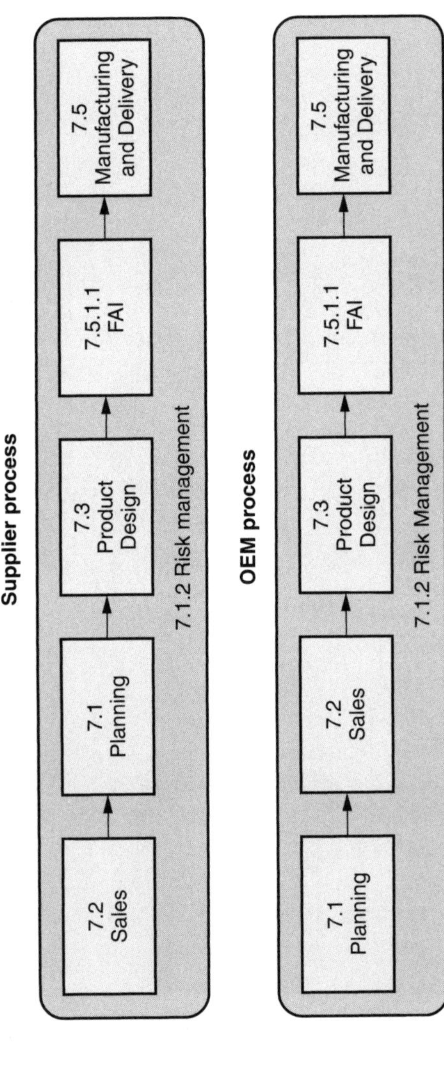

Figure 2.2 Embedding risk within the product realization process.

In case you didn't know, the FMEA methodology was originally initiated at NASA to understand the risk of launching space vehicles, and then became popular in many other industries including automotive, electronics/semiconductors, and medical devices. Only now is it once again gaining popularity within the aerospace industry.

Here are definitions of the terminology that may be common when auditing risk management:

- *Team feasibility commitment.* A commitment by the organization's product quality planning team that the design can be manufactured, assembled, tested, packaged, and shipped in sufficient quantity at an acceptable cost and on schedule. [APQP, 2nd edition]
- *Project management.* Planning and managing product realization in order to meet requirements at acceptable risk while adhering to any and all resource and/or schedule constraints.
- *Special requirements.* Those requirements identified by the customer, or determined by the organization, that have high risks to being achieved, thus requiring their inclusion in the risk management process. Factors used in the determination of special requirements include product or process complexity, past experience, and product or process maturity. Examples of special requirements include performance requirements imposed by the customer that are at the limit of the industry's capability, or requirements determined by the organization to be at the limit of its technical or process capabilities. [AS9100C]
- *Critical items.* Those items (for example, functions, parts, software, characteristics, processes) having significant effect on the product realization and use of the product, including safety, performance, form, fit, function, producibility, service life, and so on, that require specific actions to ensure they are adequately managed. Examples of critical items include safety-critical items, fracture-critical items, mission-critical items, key characteristics, and so on. [AS9100C]
- *Key characteristics.* An attribute or feature whose variation has a significant effect on product fit, form, function, performance, service life, or producibility that requires specific actions for the purpose of controlling variation. [AS9100C]
- *Design failure mode and effects analysis (DFMEA).* An analytical technique used by a design-responsible organization as a means

to assure, to the extent possible, that potential failure modes and their associated causes/mechanisms have been considered and addressed. [APQP 2nd edition]

- *(Design) verification.* Confirmation through objective evidence that specified requirements have been fulfilled. Testing to ensure that all design outputs meet requirements may include activities such as: design review, performing alternate calculations, understanding tests and demonstrations, or review of design stage documents before release. [APQP 2nd edition]
- *DVP&R.* Design verification plan and report.
- *Process flow.* The manufacturing process flow depicts the entire process, from receiving to shipping and the manufacturing processes. There are many formats of process flow used within the industry. Overall, the manufacturing process flow should be able to show the entire process from start to finish, including alternative flows. The FAI/production part approval process (PPAP) is typically based on the flow depicted in the process plan, PFMEA, and control plan of the process.
- *Process failure mode and effects analysis (PFMEA).* An analytical technique used by a manufacturing-responsible engineer/team as a means to assure that, to the extent possible, potential failure modes and their associated causes/mechanisms have been considered and addressed. [APQP 2nd edition]
- *Control plan.* Written descriptions of the system for controlling production parts or bulk materials and processes. They are written by organizations to address the important characteristics and engineering requirements of the product. Each part must have a control plan, but in many cases "family" control plans can apply to a number of parts produced using a common process. [PPAP 4th edition]
- *Enhanced FAI or PPAP.* Standardized method for use by organizations to submit approvals for new or changed parts to their customers prior to shipping mass quantities for production.

The overall integration of risk management and high-risk characteristics and their control is one of the major changes in AS9100C. There are other major and minor changes as this chapter and the next will show.

Customer Focus and Customer Satisfaction

Customer focus and customer satisfaction go hand in hand with one of the core principles of AS9101, which requires the organization to be "customer driven." According to an IAQG presentation on the changes in AS9100C, organizations need to measure and evaluate product conformity, on-time delivery performance, customer complaints, and corrective action requests. Also, the organization should develop and implement plans for improving these metrics or measurables, including customer satisfaction improvement.

Customer satisfaction is not just measuring quality or conducting customer surveys. As clause 8.2.1, Customer Satisfaction, of AS9100 states:

> Information to be monitored and used for evaluation of customer satisfaction shall include, but is not limited to, product conformity, on-time delivery performance, customer complaints, and corrective action requests. Organizations shall develop and implement plans for customer satisfaction and improvement that address deficiencies identified by these evaluations, and assess the effectiveness of the results.

Also, changes made to ISO 9001:2008 suggest that customer satisfaction needs to have multiple inputs.

Customer focus and customer satisfaction are both closely related. Customer focus is often thought of as the input into the organization while customer satisfaction is the output of how the customer feels about the organization. This relationship can be seen in Figure 2.3. Customer focus entails not just quality or delivery, but also requires understanding all the "customer needs and expectations" and implementing them within the organization. Various models can be used; see Figure 2.4 for a model

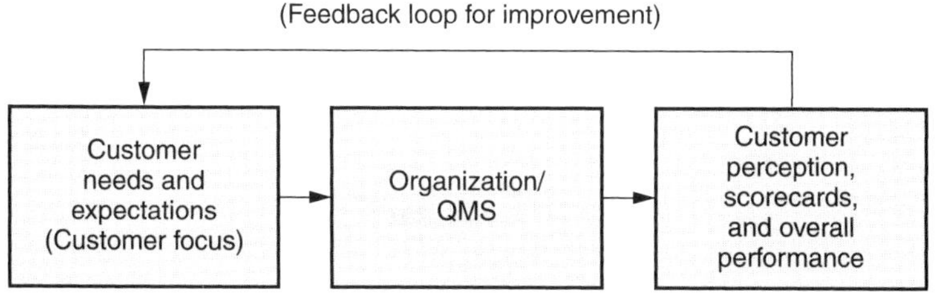

Figure 2.3 Customer focus and customer satisfaction.

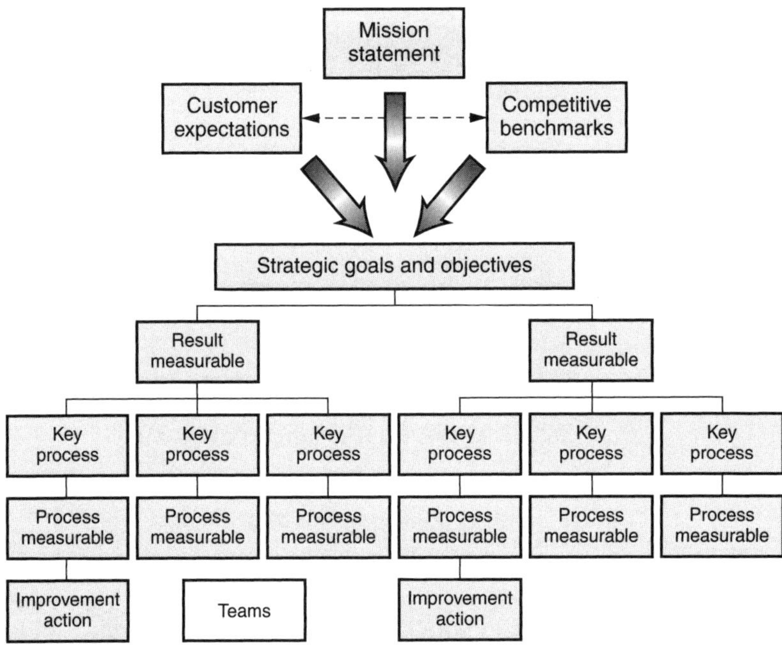

Figure 2.4 QOS/BOS model.

called QOS/BOS (quality/business operating system) that Omnex developed for Ford Motor Company and has since implemented at many organizations worldwide.

This model comprises a seven-step action-oriented model used by many organizations worldwide, from Magna to Bosch to TRW. Also, service organizations from Ryder to Kelly Services, and architectural firms such as Loaring Construction and Giffels, use this model to ensure they are customer focused and practicing continual improvement. In fact, QOS/BOS is a requirement for all Ford manufacturing and service suppliers.

AS9101 CHANGES

As mentioned earlier, the second driving force impacting the QMS is the changes to the AS9101 auditing standard. This chapter will only focus on those changes to AS9101D that affect a QMS. Auditing changes will be covered in the next chapter. AS9101 is built upon the following six core principles:

1. Customer driven—impacts the QMS
2. Process-based approach—impacts the QMS
3. Organization performance focus—impacts the QMS
4. Audit methodology—impacts the QMS
5. Quality management system effectiveness (in addition to conformity)
6. Improved reporting of audit results

Customer Core Principle 1 (CP1)—Customer Driven

This first core principle (CP1) of AS9101 requires the following:

- Customer focus (4.1.2.1): evaluation of customer satisfaction based on performance information and feedback
- Customer QMS requirements that must be audited
- Audit focused on processes that directly impact the customer
- Customer importance that must be reflected at the planned audit time
- Information and review of customer satisfaction and complaint status and processing

The key change affecting a QMS from CP1 is customer focus, as pointed out in AS9100. Both the third-party and internal audits should focus on customer focus and customer satisfaction; thus, an AS9100 QMS must have robust processes for handling customer requirements and customer satisfaction (that is, 5.2 and 8.2.1).

Though it may not be evident to the reader, the "Audit focus on the processes that directly impact the customer" is a process called *customer oriented process* (or COP), a term originating in the automotive industry. A COP is defined as a process that has an input from a customer with an output going back to the customer. Figures 2.5 and 2.6 show examples of COPs and Figure 2.7 shows an example of a Turtle diagram, terms used in automotive ISO/TS 16949:2009 auditing. The Turtle diagram is introduced as an auditing approach, but is also used by those implementing robust processes.

The next chapter will discuss these items in greater depth.

Typical implementation practices in the automotive industry include identifying the COPs and the Turtle diagrams for processes within an

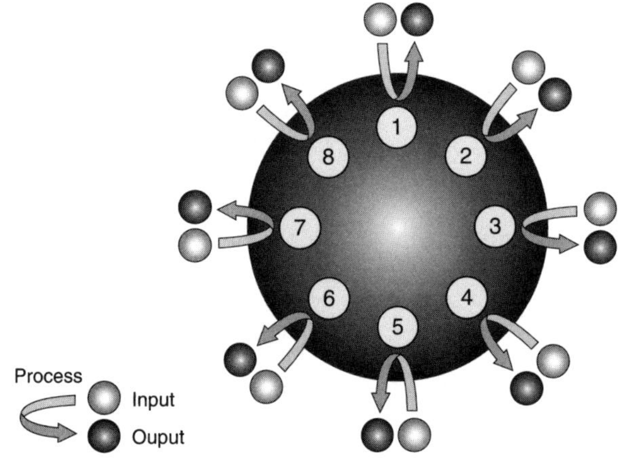

Figure 2.5 Customer oriented process example.

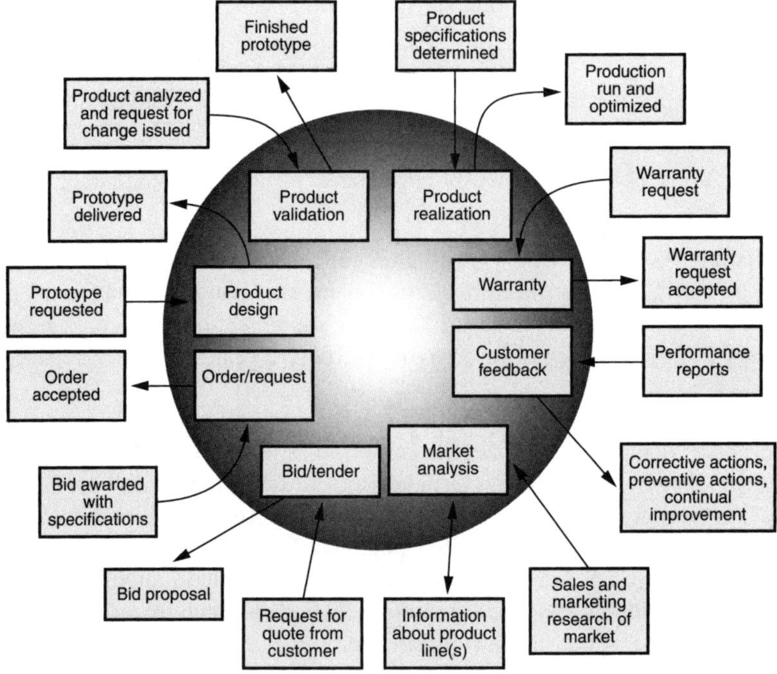

Figure 2.6 Multiple COPs in an organization, termed an *octopus diagram* by the automotive industry standards.

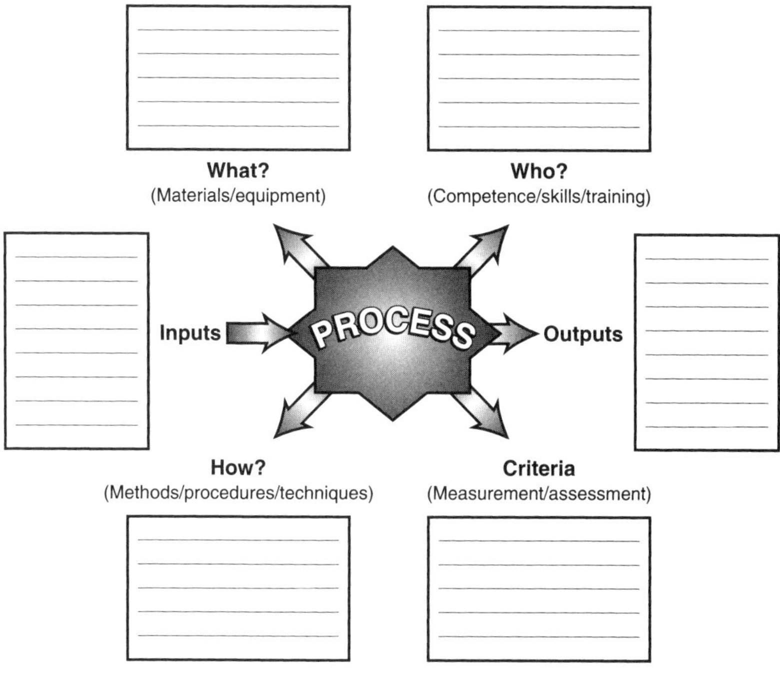

Figure 2.7 Turtle diagram example.

organization. This is an implementation practice Omnex recommends for organizations in the aerospace industry to develop their process approach.

Customer Core Principle 2 (CP2)—Process-Based Approach

The process-based approach requires the auditor to examine and audit the process used by the organization and discern whether they are following an elemental or clause approach versus a process approach. Figure 2.8 illustrates the process approach.

Many organizations in the aerospace industry have process maps that utilize a clause approach, and procedures that simply regurgitate the standard. For this reason, the process approach is viewed as a big change within the AS9100 QMS auditing and implementation scheme. The clause approach of an organization is evident in the process description, that is, process map, procedures or processes, and process measurements.

Figure 2.8 Organization process approach.

Source: Guidance on the Concept and Use of the Process Approach for Management Systems, ISO/TC 176/SC 2/N-544R3.

The process focus and customer focus can be examined/audited during the readiness review of the organization, thus allowing CP1 and CP2 to be audited/reviewed at that time.

Customer Core Principle 3 (CP3)—Organizational Performance Focus

The auditor will focus on organizational process performance, continual improvement, and customer satisfaction to determine if the process is performing effectively. Organizations preparing for this type of audit should determine process measurements and improvement goals for their primary or key processes. Omnex typically plans process performance using a business management system (BMS) control plan, which is displayed in Figure 2.9. Similarly, organizations that have conducted planning will be in a better position to answer the question, "are the processes performing satisfactorily?"

An auditor should study customer data (satisfaction, complaints, and so on), including organizational reviews, to determine how to prioritize the audit. This is conducted in the stage 1 audit.

Business Management System Control Plan

Organization ______________________ Product description ______________________

General manager ______________________ Issue/rev date ______________________

Process activity	Customer requirement/ expectation	Key process/ COP	Measurement	Responsibility	Acceptance criteria 2002	Review frequency	Control methods	Comments/ reaction
					Q1, Q2, Q3, Q4			
Examples								
Business fulfillment	On-time delivery	K	% on-time in operations	Logistics	94, 95, 96, 96%	4/yr	Monthly management meeting	Corrective action after 3 consecutive
			% on-time to customer	Production control	100%	4/yr	Trend chart	C/A is more than 15% off target
Customer complaint	Provide timely response	C	Complaint response	Quality	10 days	4/yr	Production control department	Continue to monitor
Design and development	Meet timing requirement		Time to market	Design/ development	52 weeks	Weekly	Quality department	R, Y, G reaction
Business creation	Innovation		Patents filed	Design	20 per year	Monthly	Design meeting	Continue to monitor

Figure 2.9 BMS control plan example.

Customer Core Principle 4 (CP4)—Audit Methodology

One of the audit methodologies describes process performance and effectiveness as an evaluation of process definition—including sequence and interaction, assigned responsibilities, process controls, resource availability, monitoring, measuring, and analysis—against planned results.

This description is really about the use of the Turtle diagram for auditing the process. See Figure 2.7 presented earlier. Auditors will use this tool or something similar to determine process performance and effectiveness along with process prioritization.

The use of Turtle diagrams (as taught in the IAQG Aerospace Auditor Transition Training) in turn affects QMS implementations. Typically, Omnex develops and uses certain tools to develop processes in organizations.

Customer Core Principles CP5 and CP6

The remaining customer principles, CP5, QMS effectiveness, and CP6, improved reporting of audit results, are changes to the auditing process rather than changes affecting the QMS, and hence will be covered in the next chapter.

ISO 9001:2008 CHANGES

The third major factor affecting the AS9100 QMS is the changes in ISO 9001:2008. Though these changes were introduced to the community as simply clarifications to interpretations of ISO 9001:2000, there are some changes of which auditors and those conducting implementations need to be aware.

The interpretations for this section originate from the International Register of Certified Auditors (IRCA), registrars, and other respected sources. The key changes that impact a QMS are:

- *Outsourced processes.* "'Outsourced process' is a process that the organization needs for its quality management system and which the organization chooses to have performed by an external party." Most organizations will need to evaluate those processes that they have defined as outsourced processes. Control of outsourced processes will need to be defined in the organization's QMS.
- *Inclusion of information systems in clause 6.3, Infrastructure Requirements.* AS9100C has specifically included information systems as a part of an organization's infrastructure. Organizations

will need to ensure that they are included in the description of processes, and that information system processes are defined and controlled.

- *Inclusion of recycling and disposal in clause 7.2.1, Determination of Requirements Related to the Product.* Many customers have requirements for what materials can and can not be included in the product supplied. This change, at a minimum, will require organizations to consider the recycling and disposal requirements of the customer during the customer requirements gathering process.
- *Design and development (review, verification, and validation) can be recorded together or separately.* Many organizations typically record one design review, and separate design verification and validation due to their interpretation of the standard as requiring each to be recorded separately. In most organizations, there are often many reviews, and the verification and validation activities flow together. In other words, the QMS process only documents and recognizes a small part of the real design and development process. This clarifying note in AS9100C gives organizations the freedom to abandon this practice.
- *Clause 7.3.3, Design and Development Outputs, needs to include preservation of product.* Typically, the packaging and shipping requirements are overlooked in the design output. AS9100C clause 7.3.3 requires organizations to consider the customer requirements as well as packaging and delivery when they design the product.
- *Verification of software (under AS9100C, clause 7.6).* Software used for product decisions needs to be verified both initially and on an ongoing basis. The standard also ties configuration with the need for software verification. Hence, when software configuration changes, the software needs to be reverified.
- *Additional methods for customer satisfaction (clause 8.2.1).* Customer satisfaction measurement requires multiple data inputs in AS9100C. Organizations should study all the feedback they receive from their customer and determine whether they are using all the different inputs at their organization's disposal.
- *Audit records storage (under clause 8.2.2).* Some organizations only record their nonconformities and often do not have audit checklists, audit plans, or audit reports. The standard reminds them in this update that all audit records need to be kept.

- *Guidance of measuring and monitoring of processes (clause 8.2.3).* Oftentimes, it is not clear why organizations have chosen their process measurable. This update clarifies that the organization needs to consider the type and extent of monitoring and measuring for each process to determine the impact on the conformity of product requirements and its effectiveness within the QMS.
- *Authorization of product to customer (clause 8.2.4).* AS9100C specifies that the person(s) authorizing the release of product to the customer have to be identified.
- *Reviewing effectiveness of corrective and preventive actions (clauses 8.5.2 and 8.5.3).* The standard used to simply require that the action taken is reviewed; now it requires that the effectiveness of the action taken is reviewed. Obviously, these are very different interpretations.

SUMMARY

The AS9100 QMS is influenced by changes in AS9100C, AS9101D, and ISO 9001:2008. The key changes driven by AS9100 and AS9101 are risk management, customer focus, and process focus. Risk management entails assessing the risk of product realization from contract stage to delivery of the product. Many organizations in the aerospace industry are using tools such as DFMEA, design verification plan and report (DVP&R), PFMEA, control plans, enhanced FAI, statistical process control, and measurement system analysis.

The second major change is customer focus, which, at a minimum, requires the organization to measure product conformance, on-time delivery, and customer satisfaction. There is also a requirement that processes are in place to measure performance. Many organizations implementing process focus will implement process maps and COPs, and create Turtle diagrams.

The third change to AS9100 is from ISO 9001:2008, as covered in the last section. ISO 9001:2008 and its new interpretations potentially influence an organization in a number of ways.

3

AS9110 and AS9120 Key Changes

AS9110A uses AS9100C as its base standard while AS9120A uses ISO 9001:2008 as its base standard. In other words, AS9110A includes all the AS9100C changes while AS9120A does not, but it does include all the changes made to ISO 9001:2008. As discussed in Chapter 1, AS9110 is:

> . . . intended for use by maintenance organizations whose primary business is providing maintenance, repair, and overhaul services for aviation commercial and military products; and for Original Equipment Manufacturer (OEM) organizations with maintenance, repair, and overhaul operated autonomously or that are substantially different from their manufacturing/production operations.
>
> [It] is tailored for organizations with National Airworthiness Authority (NAA) repair station certification and those that provide maintenance, repair, and overhaul services for military aviation products; but the standard could significantly benefit non-certificated maintenance organizations that choose to adopt it. *(AS9110A, clause 1.2, Application)*

The AS9120 standard, on the other hand, applies to distributors to the aviation, space, and defense industries. More specifically, AS9120 applies to:

> . . . organizations that procure parts, materials, and assemblies and resells these products to a customer in the aviation, space, and defense industries. This includes organizations that procure products and split them into smaller quantities, including those that coordinate a customer controlled service on the product. This standard is not intended for organizations that maintain or repair products. *(AS9120A, clause 1.2, Application)*

Since AS9120 is a standard for distributors, it excludes ISO 9001:2008 clauses 7.3, Design and Development, and 7.5.2, Validation of Processes for Production and Service Provision.

This chapter will cover the key changes in AS9110A and AS9120A. Keep in mind that beyond the differences in the purpose and scope of the standards, another key difference is the fact that AS9110A includes all of the AS9100C changes whereas AS9120A does not.

AS9110 REVISION A—REQUIREMENTS FOR AVIATION MAINTENANCE ORGANIZATIONS

In addition to the incorporation of changes to AS9100, AS9110 contains nine additions and one revision. This chapter will only address the incremental changes over and above AS9100. Readers who are interested in understanding the magnitude of the changes will need to read Chapter 2 as well.

Key changes in AS9110 (in addition to identified changes in Chapter 2 for AS9100):

- Clause 3.3, Counterfeit Part, and clause 3.12, Suspect Unapproved Part
- Clause 3.5, Human Factors
- Clause 8.4, Analysis of Data, and clauses 8.5.2 and 8.5.3, Corrective/Preventive Action (inclusion of Human Factors)
- Clause 3.10, Safety Policy
- Clauses 4.2.1, 5.1, 5.4.3, 5.6.1, 5.7, and 7.1, Safety Policy and Safety Objectives
- Clause 4.2.2, Quality Manual
- Clause 5.6, Management Review
- Clause 6, Resource Management
- Clause 7.5.1, Control of Production and Service Provision

The changes made to AS9110 in order to accommodate AS9100C, as discussed in the last chapter, center around the areas of risk management, customer focus, and process focus. AS9110 goes one step further with the inclusion of human and safety factors to the QMS.

Human Factors and Safety

The inclusion of human factors in clauses 8.4, Analysis of Data, and 8.5.2 and 8.5.3, Corrective/Preventive Action, more or less requires AS9110 organizations to conduct an analysis of human factors and take corrective or preventive actions based on this analysis. Also, in clause 6.2.2 there is a requirement for personnel to be trained on an ongoing basis in human factors.

Similarly, there is a running theme of safety in the standard, starting with setting policy, followed with objectives and then review. (There is also a requirement for management commitment to safety in clause 5.1):

- Clause 4.2.1.e—documented statements of a safety policy/objectives
- Clause 5.1.f/g—management commitment to safety policy/objectives
- Clause 5.4.3—management to ensure that safety objectives are established at relevant functions/levels within the organization
- Clause 5.6.1—safety policy/objectives linked to "Management Review"
- Clause 5.7—safety policy requirements further defined
- Clause 7.1.g—safety objectives and requirements for the product integrated into planning of product realization

Auditors will need to look for systems to support human factors, including training, analysis, and improvement. Likewise, safety has a more far-reaching impact, starting with commitment to implementing the objectives. Also of importance is the inclusion of safety objectives in the new product planning process.

Resource Management

The second running theme in the AS9110 changes is resource management and the overall thought that managing resources is key to maintenance organizations. The AS9110 changes affect many requirements of clause 6.0 as follows:

- Clause 6.1—system required to continually assess the availability of tools, technical data, and necessary qualified personnel

- Clause 6.2.1—personnel management expanded to ensure that certified personnel continue to maintain eligibility and that processes required for qualification and surveillance of non-certified personnel are maintained
- Clause 6.2.2—establish a training program (initial and recurrent), including personnel training associated with human factors and relevant authority and customer contract requirements

The auditor will need to study the organization's processes to manage resources including infrastructure such as tools or technical data, and resources such as qualified personnel, including competency and training programs for the same.

Counterfeit Part and Suspect Unapproved Part

Definitions were added to clearly define counterfeit and suspect unapproved parts. *Counterfeit parts* are defined as *"an article produced or altered to imitate or resemble an 'approved article' without authority or right to do so, with the intent to mislead or defraud by passing the imitation as original or genuine."* [AS9110, clause 3.3]

While it is fairly easy to understand what a counterfeit part is by its name, more uncertain is the *suspect unapproved part,* which is defined as *"an article that might not have been or is suspected of not having been produced or maintained in accordance with applicable laws and regulations. This includes: (1) parts shipped to an end user by a supplier who does not have direct delivery authorization from the approved production organization; (2) new parts that do not conform to the approved design/data; and (3) parts that have been intentionally misrepresented, including counterfeit parts."* [AS9110, clause 3.12]

Counterfeit parts and unapproved parts are a major concern to aerospace organizations, and also to the U.S. and world economy as a whole. Auditors are asked to "evaluate processes and measures taken to identify and contain counterfeit or suspected unapproved parts." In other words, not only should there be a process to ensure that counterfeit and suspected unapproved parts do not reach the maintenance spares, but also that this process needs to be audited with objective evidence during the audit process.

Quality Manual and Procedures

AS9110 requires that the quality manual include a description of maintenance processes and procedures as applicable:

- Establishing and maintaining proficiency of personnel
- Establishing and maintaining rosters for certifying staff/personnel
- Establishing and maintaining the training program
- Establishing and maintaining current approved technical data
- Performing preliminary inspection of all articles that are maintained
- Acceptance of incoming articles
- Inspecting all articles that have been involved in an accident for hidden damage before maintenance, repair, and overhaul is performed
- Conducting the maintenance process in compliance with customer, statutory, and regulatory requirements
- Performing final inspection and "return to service" of maintained articles
- Governing work performed at another location

AS9110 included these procedures in order to require all maintenance organizations to have a minimum set of basic procedures. Please note that these requirements are in addition to the basic requirements of clause 4.2.2, Quality Manual.

Management Review

Management review is impacted by the major themes running through the AS9110 changes such as customer focus, safety, and resource management. The management review needs to cover customer audits and corrective actions, and any changes from the different authorities that govern maintenance organizations. In addition, training programs and safety objectives need to be reviewed.

Specifically, the management review changes include:

- *"assessing opportunities for improvement and need for changes to the safety policy and safety objectives"* (clause 5.6.1)
- Customer audit results and requests for corrective action

- *"the achievement, adequacy, and effectiveness of the personnel training program"* (clause 5.6.2 i)
- *"changes to Authority requirements that could impact the organization"* (clause 5.6.2 j)

Control of Production and Service Provision

A few more process controls were added to the process of conducting maintenance:

- Clause 7.5.1 h—maintenance and inspection/verification meets the technical data as *"approved . . . or that is acceptable to the Authority"*
- Clause 7.5.1 k—*"Criteria for workmanship, specified in the clearest practical way . . . in accordance with applicable technical data"*
- Clause 7.5.1 o—*"the equipment, tools, and materials shall be those recommended by the manufacturer of the article . . . and acceptable to the customer and/or Authority."*

Some of the changes represent some level of effort, such as ensuring that maintenance and inspection/verification are being conducted as planned, and the criteria for workmanship for maintenance activities. The standard suggests using written standards, representative samples, and illustrations.

AS9120A—REQUIREMENTS FOR AVIATION, SPACE, AND DEFENSE DISTRIBUTORS

AS9120A—unlike AS9110A—does not include all the changes in AS9100, but only includes six additions including risk management, planning for product realization, configuration management, and other changes, including five revisions/relocations and two deletions. The changes are over and above the changes to the ISO 9001:2008 standard.

Counterfeit and suspect parts are already covered in the AS9110 standard and are repeated again in AS9120. Planning for product realization, configuration management, and work transfer, though new to both AS9110A and AS9120A, have always been a part of AS9100 and really don't need detailed explanations.

The key changes to AS9120 include the following.

Clauses 3.3 and 3.7—Counterfeit Part/Suspected Unapproved Part (SUP)

This requirement for distributors is the same as it is for maintenance organizations. Similarly to AS9110, definitions were added to clearly define counterfeit and suspect unapproved parts. A counterfeit part is defined as *"a product produced or altered to imitate or resemble a product without authority or right to do so, with the intent to mislead or defraud by passing the imitation as original or genuine."* [AS9120, clause 3.3]

While it is fairly easy to understand what a counterfeit part is by its name, more uncertain is the suspect unapproved part, which is defined as *"a product that might not have been or is suspected of not having been produced or maintained in accordance with applicable laws and regulations."* [AS9110, clause 3.7]

Counterfeit parts and unapproved parts are a major concern to aerospace organizations and also to the U.S. economy as a whole. Auditors are asked to "evaluate processes and measures taken to identify and contain counterfeit or suspected unapproved parts." In other words, not only should there be a process to ensure that counterfeit and suspected unapproved parts do not reach the maintenance spares, but also that this process needs to be audited with objective evidence during the audit process.

The definitions, actions to take, and audit points are the same as for AS9110 above.

7.1—Planning of Product Realization

Planning was excluded for distributors in the previous revision of the AS9120 standard. However, it is a requirement (as it should be) in the latest revision of the standard. In other words, when the distributor is introducing new products to existing customers, or existing products to new customers, and/or new products to new customers, there must be a planning process to ensure that the service provided has process control (clause 7.5) and is packaged and delivered on time with high quality.

Distributors need to consider all the requirements of AS9101, including risk management, when planning new products. Auditors need to evaluate the plannning process for AS9120 organizations.

7.1.1/7.5.3—Configuration Management

Configuration management was added to the AS9120 standard. This includes configuration planning, identification, change control, accounting, and auditing. Auditors will need to audit for this process.

7.1.2—Work Transfer

Work transfer refers to a requirement for the distributor to establish, implement, and maintain a process to plan and control temporary or permanent transfer of work within one organization or from one supplier to another.

SUMMARY—AS9110 AND AS9120 CHANGES

AS9110 includes all the AS9100 changes, including customer focus, process focus and risk management, and the ISO 9001:2008 changes. Additionally, maintenance organizations need to include processes for human factors, safety, resource management, and other changes.

AS9120, on the other hand, does not include all the changes in AS9100, but only includes six additions, including risk management, planning for product realization, configuration management, and other changes. Organizations subscribing to AS9120 will also need to include the relevant ISO 9001:2008 changes.

4

Understanding AS9101D Auditing Requirements— What Has Changed

Simply put, AS9101D is *a complete departure* from the previous AS9101 standard. So what has changed? Essentially, everything. The AS9101 writing committee practically threw away the previous standard and started from scratch when they wrote the AS9101D revision.

For starters, AS9101 now acts as a single, unified auditing standard for AS9100, AS9110, and AS9120. The highlights of the standard include:

- An enhanced audit process for evaluating "process-based management systems"
- Prioritization of the audit based on customer and performance data
- Auditing process effectiveness and process conformance
- Focus placed on process performance

AS9101D was authored by 13 members on the 9101 team—from six different countries representing the Americas, Europe, and Asia-Pacific IAQG sectors—which included eight IAQG member companies, one certification body, and four members from the stakeholder community.

The AS9101 auditing process can be summarized by Figure 4.1. AS9101D is laid out as follows.

Audit methodology:

- Clause 4.1.2.1, Customer Focus
- Clause 4.1.2.2, Organizational Leadership
- Clause 4.1.2.3, Quality Management System Performance and Effectiveness
- Clause 4.1.2.4, Process Management

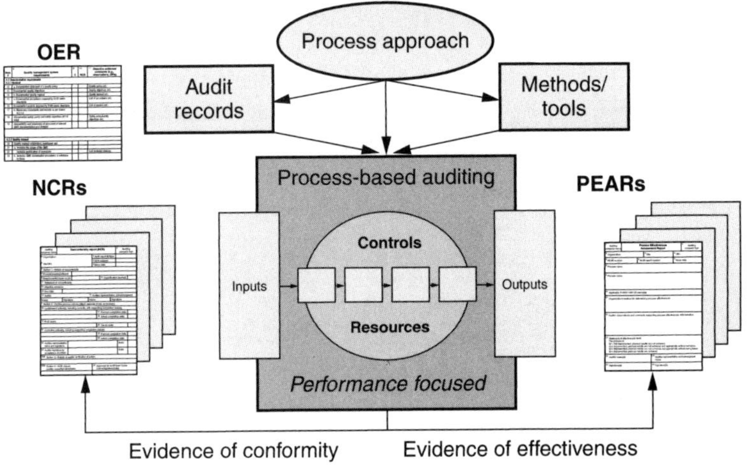

Figure 4.1 Summary diagram of AS9101 auditing process.

Source: AS9101 model for process-based auditing, IAQG 9101:2009 change overview presentation, April 2010.

- Clause 4.1.2.5, Process Performance and Effectiveness
- Clause 4.1.2.6, Continual Improvement

This is followed by common audit activities:

- Clause 4.2.1, Audit Planning
- Clause 4.2.2, On-Site Audits
- Clause 4.2.3, Audit Reporting
- Clause 4.2.4, Nonconformity Management
- Clause 4.2.5, Audit Records

Finally comes the audit phases:

- Clause 4.3.1, Pre-audit Activities
- Clause 4.3.2, Stage 1 Audit
- Clause 4.3.3, Stage 2 Audit
- Clause 4.3.4, Surveillance
- Clause 4.3.5, Recertification

In order to discuss what is new in AS9101D, it is best to start with the process approach itself, then follow up with audit methodology, common audit activities, and finally, audit phases. When discussing the conducting of the internal audit for AS9100, AS9110, and AS9120, this book will take what is best from the third-party auditing process described in AS9101D. Conducting formal internal audits using an approach similar to third-party auditors works best for internal auditors conducting quality system audits.

PROCESS APPROACH

This may be the biggest change to the audit approach in AS9100, AS9110, and AS9120. In its introduction, AS9101D lists some basic questions that need to be asked of each process:

1. Is the process identified and adequately defined?
2. Are responsibilities defined?
3. Are processes implemented and maintained?
4. Is the process effective in achieving desired results?

Nowhere is this process approach more evident than in clauses 4.1.2.4, Process Management, and 4.1.2.5, Process Performance and Effectiveness.

Clause 4.1.2.5, Process Performance and Effectiveness, includes the assessment of process control, objectives/targets (process effectiveness), and continual improvement, all of which can lead to customer-related issues. The audit plan will be prioritized during stage 1 to consider customer-related issues, and the Process Effectiveness Assessment Report (PEAR) audit report tool can be used to evaluate the effectiveness of the process. Additionally, the auditor can use the Turtle diagram process analysis tool to help plan and conduct effective interviews that address the inputs and outputs of a process, process measurables, resources, training, and standards/documentation for the process. Figure 4.2 is an example of a blank PEAR, and Figure 4.3 displays a Turtle diagram.

Collectively, per AS9101D, the tools must assess the following:

- Process responsibilities assigned and responsible functions identified (process name and owner)
- Process activities
- Process limits (inputs and outputs)
- Process control methods

1 *Auditing company name*	**Process Effectiveness Assessment Report**		2 *Auditing company logo*
3 Organization:		4 Site:	4 OIN:
5 PEAR number:	6 Audit report number:		7 Issue date:
8 Process name:			
9 Process details, including associated process interfaces:			
10 Applicable 9100/9110/9120 clause(s):			
11 Organization's method for determining process effectiveness:			
12 Auditor observations and comments supporting process effectiveness determination:			
13 Statement of effectiveness level: The process is: ❑ 1. Not implemented; planned results are not achieved. ❑ 2. Implemented; planned results are not achieved and appropriate actions not taken. ❑ 3. Implemented; planned results are not achieved, but appropriate actions being taken. ❑ 4. Implemented; planned results are achieved.			
14 Auditor name(s):		15 Auditee representative acknowledgment name:	
14 Signature(s):		15 Signature(s):	

Figure 4.2 Process Effectiveness Assessment Report (PEAR) example.

- Process operation methods
- Availability of resources and information, including training and competency
- Applicable AS9100 requirements
- Applicable customer, regulatory, and statutory QMS requirements

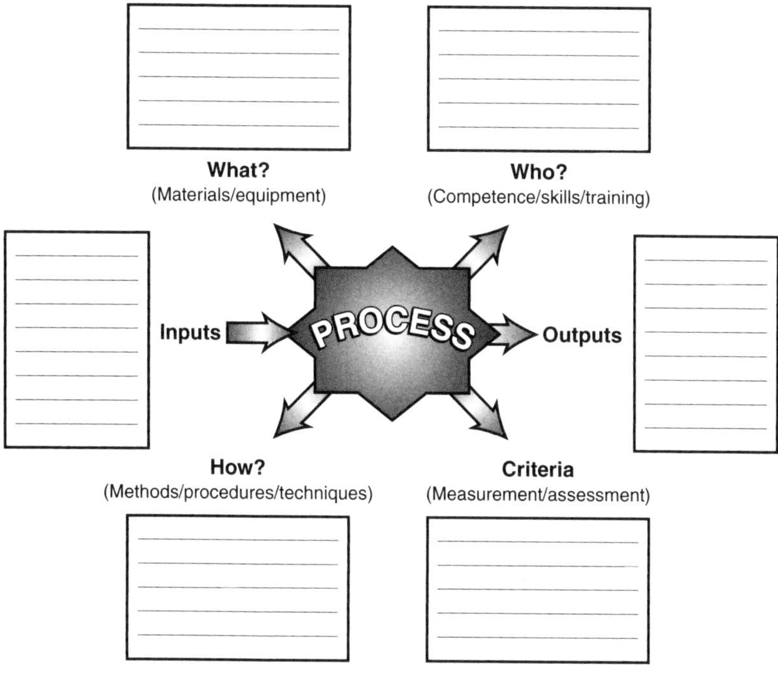

Figure 4.3 Turtle diagram example.

- Whether processes are monitored, measured, and analyzed against planned results
- Actions implemented to achieve planned results
- Process effectiveness in achieving desired result

AUDIT METHODOLOGY

Clauses 4.1.2.1, Customer Focus, and 4.1.2.2, Organizational Focus

AS9101D requires an organization to have a customer feedback process. This process is key to customer satisfaction and should be one of the prioritized processes in the audit plan developed during stage 1 for the stage 2 audit. At the same time, the different customer inputs such as customer satisfaction, on-time delivery, and product quality, including customer complaints (see AS9101D for a longer list), should be studied and identified in the audit plan, with the audit trails carefully chosen based on processes that either positively or negatively impact customer satisfaction.

The process should include customer inputs from the different buying centers of the customers in aerospace, space, and defense industries. The auditor should study how the process is managed, how data are collected, and how well the organization studies the data gathered and responds to customer-related issues through actions and decisions.

Another key AS9101D audit requirement is the interview of top management of the organization. AS9101D clause 4.1.2.2, Organizational Leadership, suggests that the interview assess the following:

- Establishment and continued relevance of the quality policy and objectives
- Establishment of performance measures aligned to quality objectives
- Quality management system development, implementation, and continual improvement
- Top management commitment
- Quality management system performance and effectiveness
- Performance to customer expectations (for example, supplier rating, scorecard, audit results)
- Actions taken to address issues that are not meeting customer performance expectations

Clause 4.1.2.3, Quality Management System Performance and Effectiveness

This audit method evaluates the effectiveness of the quality management system by studying a number of measures, including:

- Handling of customer complaints, feedback, and other relevant customer data
- Results and actions of previous audits (both internal and third-party), including associated records
- Stakeholder feedback, such as from regulatory authorities or other interested parties

- Handling of process and/or product nonconformities, and a review of the corrective actions and an evaluation of their effectiveness
- Handling of preventive actions, and an evaluation of the actions taken
- Conduct of management review, including a review of all associated records
- Internal performance monitoring, measurement, reporting, and reviews against performance objectives and targets, and a review of continual improvement activities and associated records
- Performance against targets, and review of records of corrective actions where targets are not being met
- Status and effectiveness of the process performance improvement activities and the outcomes related to product quality

Finally, effectiveness is measured by whether the organization is meeting customer satisfaction and organizational quality objectives.

Clauses 4.1.2.4, Process Management, and 4.1.2.5, Process Performance and Effectiveness

Clause 4.1.2.4, Process Management, suggests that the auditor review the sequence and interaction of processes, with a focus on processes impacting the customer. In short, "sequence and interaction" is typically documented using a process map. An example of a process map is given in Figure 4.4. This means the auditor must be able to evaluate a process map, focusing on those processes important to the customer, that is, the customer oriented processes (COPs) that we defined in Chapter 2 (Figures 2.5 and 2.6).

Clause 4.1.2.6, Continual Improvement

Continual improvement is a continuation of the prioritization of the audit, as conducted for customer focus. The auditor chooses audit trails to *"ensure focus on issues that are important to the organization, their customers, and regulatory authorities."* AS9101D suggests the development of audit trails that cover the following activities:

- Quality policy and objectives used to reflect commitment
- Identified improvement actions using various inputs, such as:

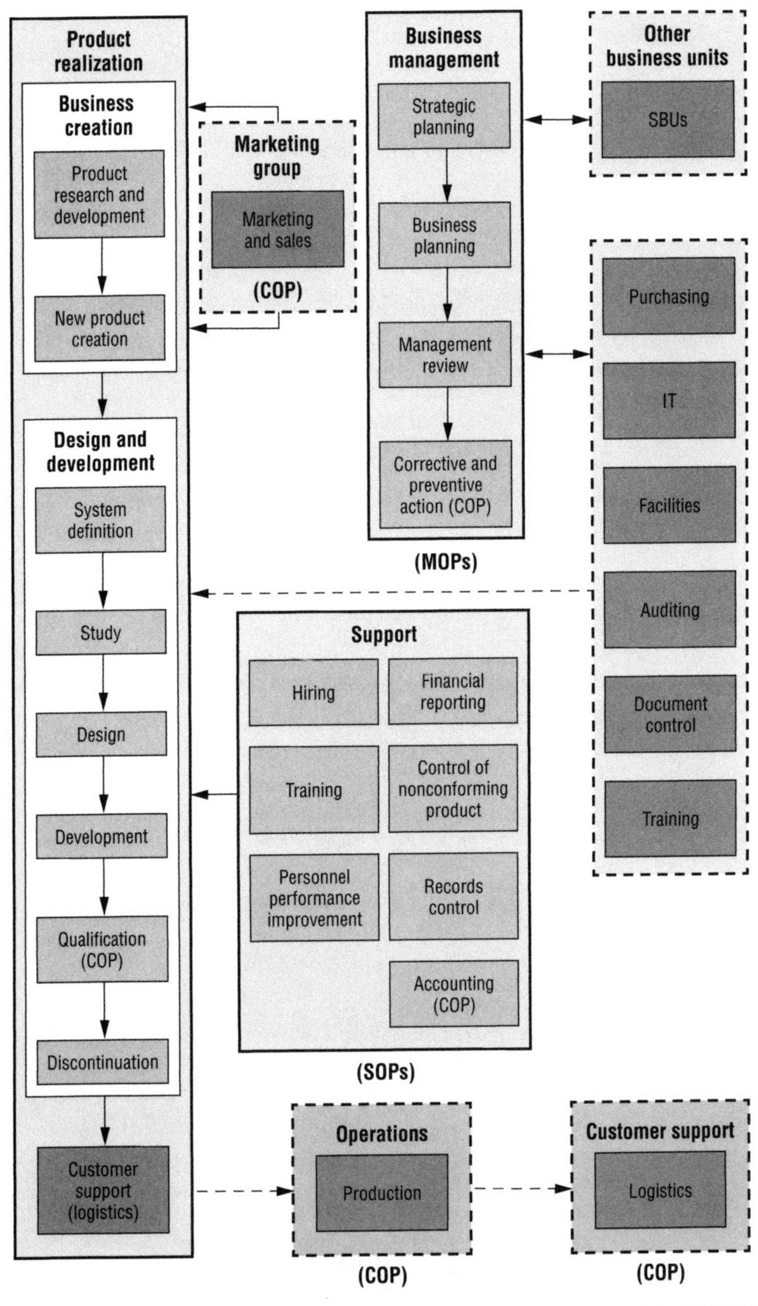

Figure 4.4 Process map example.

  - Results of internal and external audits
  - Customer satisfaction data, including product quality, on-time delivery performance, and complaints
  - Process performance
  - Aggregate information on corrective and preventive actions
  - Problem resolutions and lessons learned
  - Benchmarking and sharing of best practices
- Management review used to evaluate the status of an improvement action
- Availability and management of resources
- Continual improvement implementation plans
- Monitoring and measurement of continual improvement activities, including processes for:
  - Monitoring and measurement of continual improvement actions and results
  - Evaluating effectiveness of actions taken and associated results
  - Initiating modified/additional actions if results are not satisfactory or effective

COMMON AUDIT ACTIVITIES

The common audit activities of AS9101D (clause 4.2) as they relate to the audit phases (clause 4.3) are shown in Figure 4.5.

In order to ensure that the management system is operating effectively, Omnex recommends both quality system audits and process/product audits. This book covers only internal audits of the quality management system.

For inexperienced auditors, or auditors lacking a background in ISO 19011, it may be beneficial to attend an internal auditor and/or lead auditor training class in order to understand the steps of ISO 19011, including stage 1 (including audit planning) and stage 2 (including opening meeting, writing nonconformities, and closing meeting) and corrective action close-outs. This section will focus on those aspects of aerospace auditing over and above the ISO 19011 auditing standard. In fact, AS9101D auditing includes ISO 19011, ISO 17021, and additional aerospace requirements specific to aerospace auditing relevant for third-party audits. Those aerospace-specific auditing requirements will be covered in the next chapter.

Common activity (4.2) \ Audit phase (4.3)	Pre-audit activities (4.3.1)	Audit program			
		Stage 1 (4.3.2)	Stage 2 (4.3.3)	Surveillance (4.3.4)	Recertification (4.3.5)
Audit planning (4.2.1)	✓	✓	✓	✓	✓
On-site auditing (4.2.2)		✓	✓	✓	✓
Audit reporting (4.2.3)		✓	✓	✓	✓
Nonconformity management (4.2.4)			✓	✓	✓

Figure 4.5 Relationships between audit phases and common audit activities.

Source: AS9101D standard.

Clause 4.2.1, Audit Planning

Audit planning needs to focus on the processes of the organization, and should be developed based on the priorities to the customer and the organization. Audit trails need to be developed that consider these priorities, such as: criticality of products and processes, including special processes; product safety (airworthiness issues); results of internal audits; previous audit findings; performance measures and trends for quality and on-time delivery (OTD), management review results; customer satisfaction; and complaints; including OASIS reports. As every organization is unique, so too are their priorities. The auditor must consider these unique priorities when developing audit trails.

Clause 4.2.2, On-Site Auditing

On-site audits need to include the following:

- Review of changes to the QMS since the last audit
- Review of requirements for new customers
- Review of customer satisfaction
- Interview of top management
- Assessment of quality management system performance and effectiveness
- Other items as applicable

Identifying and Recording of Audit Findings

Third-party auditors are required to record objective evidence in a standardized form called the Objective Evidence Record (OER), an excerpt of which can be seen in Figure 4.6. For internal auditors, this book offers an audit checklist with guidance on conducting effective audits in Appendix A.

Item #	8 **Quality management system requirements**	9 **C**	10 **NCR**	11 **Objective evidence/ comments (e.g., observations, OFIs)**
4.2 Documentation requirements **4.2.1 General**				
12	a. Documented statement of a quality policy			Quality policy ref.:
13	Documented quality objectives			Quality objectives ref.:
14	b. Documented quality manual			Quality manual ref.:
15	c. Documented procedures required by 9100-series standards			List of procedures ref.:
16	Documented records required by 9100-series standards			List of records ref.:
17	d. Necessary documents and records as per clause 4.2.1.d			
18	*Documented safety policy and safety objectives (9110 only)*			Safety policy/safety objectives ref.:
19	*Assessibility and awareness of personnel of relevent QMS documentation and changes*			
4.2.2 Quality manual				
20	Quality manual established, maintained and			
21	a. Includes the scope of the QMS			
22	Includes justification of exclusions			List excluded clauses:
23	b. Includes QMS documented procedures or reference to them			
24	c. Includes a description of the QMS processes and interactions			
	d. Includes a description of the processes and procedures use for: (9110 only)			
25	• *proficiency of personnel*			
26	• *rosters of certifying staff and personnel*			
27	• *training program*			
28	• *current approved technical data*			
29	• *performing preliminary inspection*			
30	• *acceptance of incoming articles*			
31	• *inspecting articles involved in an accident for hidden damage before MRO activities (9110 only)*			
32	• *conducting maintenance in compliance with customer, statutory, and regulatory requirements*			
33	• *performing final inspection and "return to service" of maintained articles*			
34	• *governing work performed at another location*			

Figure 4.6 Objective Evidence Record excerpt.

AS9101D also offers a standardized Nonconformity Report (NCR) to document nonconformities, as seen in Figure 4.7. This same form is recommended for internal auditors. A few rules that apply to third-party auditors will work for internal auditors, such as the need for the recurrence of the same or similar nonconformity found during consecutive audits in a site/location to be classified as a major nonconformity. Also helpful to internal auditors is discerning between nonconformities and observations.

1 *Auditing company name*	**Nonconformity Report (NCR)**		2 *Auditing company logo*
3 Organization:		4 Audit report ID/type	
		4 NCR number:	
5 Site/OIN:		6 Issue date:	
7 Section 1—Details of nonconformity			
8 Process/area/department:			
9 Requirements/clause no.(s):		10 Classification (ma/mi):	
11 Statement of nonconformity:			
12 Objective evidence:			
13 Due date:			
14 Auditor		15 Auditee representative acknowledgment	
Name:	Signature:	Name:	Signature:
7 Section 2—Auditee planned actions (Attach separate sheet, as needed.)			
16 Containment action(s), including correction, with supporting completion date(s):			
		17 Planned completion date:	
		18 Actual completion date:	
19 Root cause:			
		20 Cause code:	
21 Corrective action(s), including supporting completion date(s):			
		22 Planned completion date:	
		23 Actual completion date:	
15 Auditee representative name and signature:		Date:	
24 Auditor signature for acceptance of C/A(s):		Date:	
7/25 Section 3—Details of auditor verification of action:			
7/26 Section 4—NCR closure (auditor name/signature/date):		27 Approved by audit team leader (name/signature/date):	

Figure 4.7 Standard Nonconformity Report example.

Internal auditors should keep in mind that the standard definition of a *non-conformity* is "non-fulfillment of a requirement," and it can not be graded as an observation, opportunity for improvement, or recommendation.

Lastly, process effectiveness is documented using a PEAR report as discussed earlier in this chapter (see Figure 4.2). This form is also beneficial to internal auditors and will be covered in detail in Chapter 8, Stage 2: On-Site Audit.

Special Processes

Special processes need to be evaluated for process validation and monitoring, measurement, and control as identified in the audit plan. The additional requirements of AS9101D that are added to the internal auditor checklist are to ensure that the special processes meet customer requirements and the calibration records of the monitoring and measuring equipment. The checklist also can be used to check the requirement that the traceability between the product and the process can be verified. For outsourced special processes, the control of the supplier and use of customer-designated sources should be checked.

Clause 4.2.3, Audit Reporting

AS9101D requires the use of a Process Matrix Report, an excerpt of which can be seen in Figure 4.8. The Internal Audit Report provided in this book will also use this matrix in order to help track processes planned for the audit in stage 1 and what was finally audited in stage 2. It will help keep track of processes audited between system audits. AS9101D provides an audit report for both stage 1 and 2. An internal audit report version of both of these reports is provided in Appendixes A and B.

Stage 1 report content:

a. Organizational details

b. Audit criteria

c. Business and customer breakdown

d. High-level requirements confirmation (quality manual and required documented procedures)

e. Key customer performance quality and delivery

f. Customer quality management system approval status

g. Additional aviation, space, and defense quality management system requirements

1 *Auditing company name*	**QMS Process Matrix Report**										2 *Auditing company logo*
3 Organization:						5 Audit report number:					
4 Site/OIN:						6 Issue date:					
7 Standard: 9100 ❑ 9110 ❑ 9120 ❑											
	Company QMS processes										
	1	2	3	4	5	6	7	8	9	10	
8 Process name											
9 Related Process Effectiveness Assessment Report (PEAR ID)											
Clauses * = not applicable for 9120	10 Conformity										11 NCR number
	1	2	3	4	5	6	7	8	9	10	
4.1 General requirements											
4.2.1 Documentation requirements—General											
4.2.2 Quality manual											
4.2.3 Control of documents											
4.2.4 Control of records											
5.1 Management commitment											
5.2 Customer focus											
5.3 Quality policy											
5.4.1 Quality objectives											

Figure 4.8 Process Matrix Report example.

h. Key information (specific information obtained from organization)

i. Audit team leader recommendations (readiness for stage 2)

j. Organization confirmation

k. Signature

Stage 2 report content:

a. Organizational details

b. Audit scope and objectives

c. Audit team

d. Nonconformity summary

e. Summary of PEAR reports

f. Audit summary

g. Key issues/concerns for top management

h. Strengths and good practices

i. Opportunities for improvement/observations

j. Previous audit nonconformity status

k. Changes to the organization/facilities/quality management system/scope

l. Agreed follow-up requirements

m. Audit team leader recommendations

n. Organization confirmation

o. Signature

Clause 4.2.4, Nonconformity Management

AS9101D has specific requirements for nonconformity management that are helpful for internal audits as well, from writing the statement of nonconformity to recording objective evidence. The standard Nonconformity Report in Figure 4.7 details the nonconformity closure and verification.

Clause 4.2.5, Audit Records

Audit records include the checklists, stage 1 audit report, audit notes, and records of objective evidence.

AUDIT PHASES

Clause 4.3.1, Pre-Audit Activities

Stage 1 Information Required

- Quality manual
- Description of processes showing their sequence and interactions, including outsourced processes

- Performance measures and trends for the previous 12 months
- Evidence that the requirements of the standard are addressed by the organization's documented procedures
- Interactions with support functions on-site or at remote locations/sites
- Evidence of internal audits of processes/procedures, including internal and external quality management system requirements
- Latest management review results
- List of major aviation, space, and/or defense customers requiring AS9100 series standards
- Evidence of customer satisfaction and complaint summaries, including verification of customer reports, scorecards, and special status or equivalent

Clause 4.3.2, Stage 1 Audit—Conclusions/Outputs

- Quality management system implementation status
- Organization's readiness for stage 2 audit
- Areas of concern that could result in a nonconformity
- Audit plan for stage 2 audit
- Verify scope of certification and applicability to IAQG scheme, audit day calculations
- Audit time for stage 2
- Composition of team including technical experts and translators
- Changes to contract

Clauses 4.3.3, Stage 2 Audit, and 4.3.4, Surveillance

The stage 2 audit must cover all the clauses of the applicable standard(s) and associated processes while surveillance audits need to cover all clauses and processes of the organization within the certification cycle of three years.

All nonconformities need to be closed and verified before a recommendation for certification can be made by the lead auditor.

Repeat nonconformities, lack of performance data, or lack of operational control are all grounds to suspend the certification during the surveillance cycle.

Clauses 4.3.5, Recertification, and 4.3.6, Special Audits

Recertification audits need to be conducted three months before the expiration of the certificate. Special audits are required to be conducted when a customer or other interested party provides objective evidence of a serious issue or change in scope, or if the organization changes registrars. There is a unique audit report provided by AS9101D for special audits.

SUMMARY OF AS9101D

AS9101D consists of three different sections—Audit Methods/Tools, Common Audit Activities, and Audit Phases. In total, there are six audit methods and tools: customer focus, organizational leadership, quality management system performance and effectiveness, process management, process performance and effectiveness, and continual improvement. These audit methods are used in the common audit activities, which are audit planning, on-site auditing, audit reporting, and nonconformity management. These same activities repeat themselves in the five phases of an audit: pre-audit activities, stage 1 audit, stage 2 audit, surveillance, and recertification.

In summary, aerospace auditing is differentiated by the customer and process focus prevalent in the audit requirements. The audit is prioritized using audit trails based on customer and organizational performance. This focus will be reflected when the processes and process owners are audited during the on-site audit.

Secondly, the organization's process approach and process definition need to be audited. Questions to ask include: how are the processes planned? who is the owner? is it measured? is it improving? and if not, is there an action plan?

More on customer and process focus can be found in the next chapter—the Aerospace Auditing Approach.

5

The Aerospace Auditing Approach—Process Approach, Customer Focus, and Leadership

This chapter is a continuation of the last chapter on the AS9101D standard. Aerospace auditing has three main pieces—process approach, customer focus, and leadership auditing. The *process approach* consists of two separate parts: first to analyze an organization to ensure that it is process-oriented, and then to audit processes to ensure they are effective. *Customer focus* prioritizes the audit, identifies whether there is a customer feedback process, whether the organization is satisfying their customer needs, and whether there are action plans for improvement. *Leadership auditing* includes evaluating the knowledge and commitment of organizational leadership.

PROCESS APPROACH

The process approach audit includes analyzing an organization's process approach and determining whether selected processes are effective. The first part takes place during stage 1 and will be indicated in the stage 1 audit plan as the "process approach audit." The internal auditor will need to conduct this only during the organization's initial application for their upgrade from AS9100 Rev B to AS9100 Rev C. In subsequent audits, the internal auditor can study the process approach and continue to move their organization toward the best practices of the process approach, as this chapter will explain.

Understanding the Process Approach and the Organization's Processes

Before conducting an AS9100, AS9110, or AS9120 audit, the auditor must ensure that the organization is process-focused. The auditor's process-focus expectations are framed by at least three directives:

- Clause 4.1, AS9100, AS9110, and AS9120
- Clause 4.1.2.4, Process Management (AS9101)
- Clause 4.1.2.5, Process Performance and Effectiveness (AS9101)

Understanding a Process Map

Processes and their sequence and interaction, including outsourced processes, need to be identified per clause 4.1, General Requirements, in AS 9100/9110/9120 and clause 4.1.2.4, Process Management, in AS9101D. As AS9101D states in a clarifying note in clause 4.3.2.2, Collection of Information, "processes can be depicted in various ways [for example, process maps, Turtle diagrams, SIPOC method (breakdown of supplier, inputs, process steps/tasks, outputs, customer), octopus]." Since most organizations are going to be using process maps for this purpose, this chapter will discuss process sequence and interaction in terms of auditing a process map.

The process map is made up of process blocks that consist of groups of individual processes. For example, the product realization block would contain research and development, sales, product design, process design, manufacturing, and other related processes. To maintain this logical sequence, support process or management process blocks should be added to the process map. An important process in the management process block is how the organization conducts business strategy, objectives, and management review processes. Figure 5.1 shows an example of a process map.

The process map should be limited to high-level processes that, in turn, may involve one or more other processes that are documented in level II procedures. A one-to-five ratio in the relationship between processes in the process map and subprocesses in the procedures manual is typical. For example, the process of strategic planning may include gathering customer expectations, benchmarking, and setting and deploying objectives.

Process Hierarchy

The hierarchy of the aerospace process approach includes processes that directly impact the customer, as stated in clause 4.1.2.4a of AS9101D. In this chapter, these will be called *customer oriented processes* (COPs). Examples of COPs are shown in Figure 5.2. It is not important for the organization to identify COPs in their process map, but only that the auditor focuses on them during the evaluation of the process map. When evaluating the process map, the auditor can study whether the customer-oriented

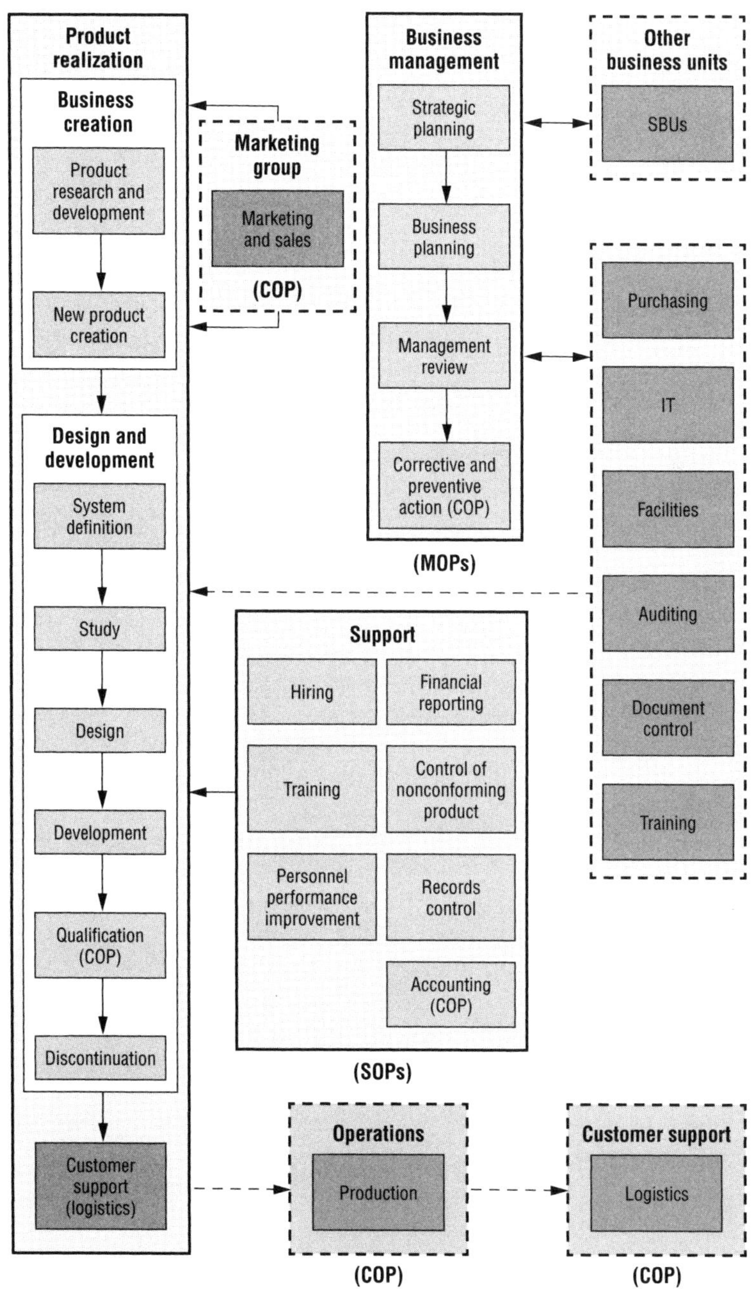

Figure 5.1 Process map example.

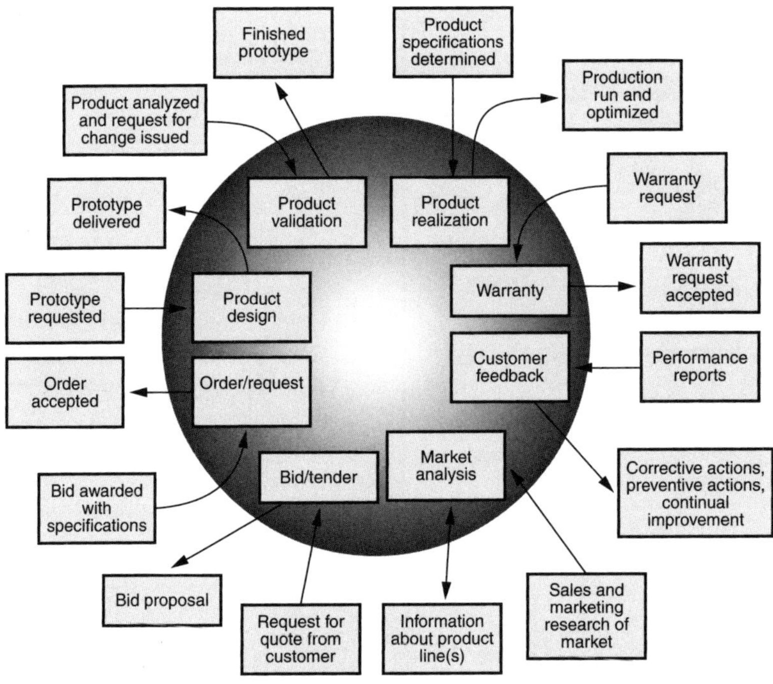

Figure 5.2 Multiple COPs in an organization, termed an *octopus diagram* by the automotive industry standards.

processes are present and then determine which ones will be selected to target for the process effectiveness audit using the PEAR form during the stage 2 on-site audit. (*Note:* It is mandatory for third-party auditors to use the PEAR form to evaluate the product realization process.)

COPs are not specifically referenced by the AS9100 series of standards. AS9101D only mentions the importance of processes that directly impact the customer and, as such, COPs fulfill that definition.

Examples of Customer Oriented Processes

- Market analysis
- Bid and/or tender
- Order and/or request

- Product and process verification and/or validation
- Product production
- Delivery
- Payment
- Warranty and/or service
- Customer feedback

A little bit more about COPs: COPs, by definition, receive input from the customer and send output back to the customer. The auditor should study the process map to ensure that the input and output are shown clearly for the COPs in both the process map and the associated process documentation.

Process maps typically should have at least a "product realization" or "new product launch" block with processes that start with sales or R&D and continue into manufacturing and then finally into shipping the product to the customer. There should also be a block that manages and governs the organization, including the setting of strategy and conducting business/management reviews of the system. There is also a third block of support processes that support the entire organization. Figure 5.3 displays these blocks.

Many linkages are possible between processes. If the auditor expects to see all the linkages on the process map, it would look like a tangle of spaghetti, and the resulting information would be meaningless. One has to balance the number of linkages shown with adding value to the process map and the information it is trying to convey.

Of course there should be links, though none of the links are shown in Figure 5.3.

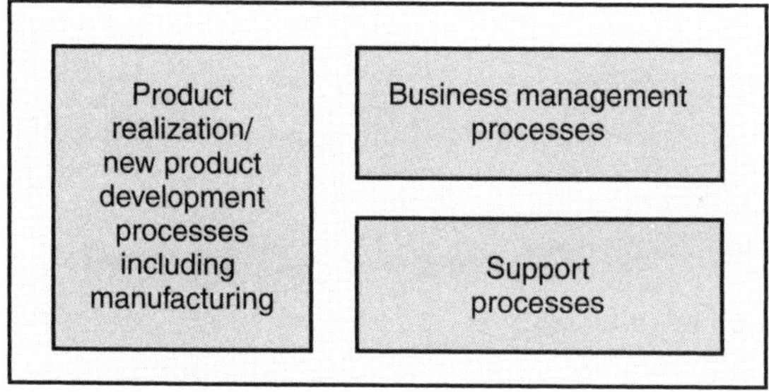

Figure 5.3 Process map blocks and expectations of an auditor.

Process Interfaces between Sites in an Organization

An organization is made up of many different sites that are managed by different units connected by an organizational hierarchy. There are, for example, corporate headquarters, corporate sales, purchasing, and design, with multiple plants and multiple sales offices. How are these organizational units linked? Are they linked effectively by processes that flow between them? Does the process that sets overall company objectives start at corporate headquarters, and does the process show how they are deployed and established at the manufacturing site and throughout the enterprise? These are some of the questions auditors must consider as they evaluate an entity.

The scope statement in a quality manual must include the functions and organizational units, whether on-site or remote. The process map goes hand in hand with the entities of a company. Corporate sites, manufacturing sites, and supporting locations share processes and a flow of information between them. The "process interface" referred to in AS9101D is the links, communication, and information flows between the different sites. Thus, when a sales process is studied at corporate headquarters, does it clearly show how it interacts with the product design and manufacturing plant? Does it show how the product risk is determined?

The process map should illustrate the links between the sites and support locations for many processes, including:

- Sales
- Purchasing
- Objectives deployment
- Management review data
- Management commitment
- Human resource planning
- Program management
- Advanced product quality planning (APQP) or new product development
- Customer complaints

The process map can not stop within the four walls of the remote location or the manufacturing site. It has to explain the flow of information and action between the site and the remote location. This is demonstrated in Figure 5.4.

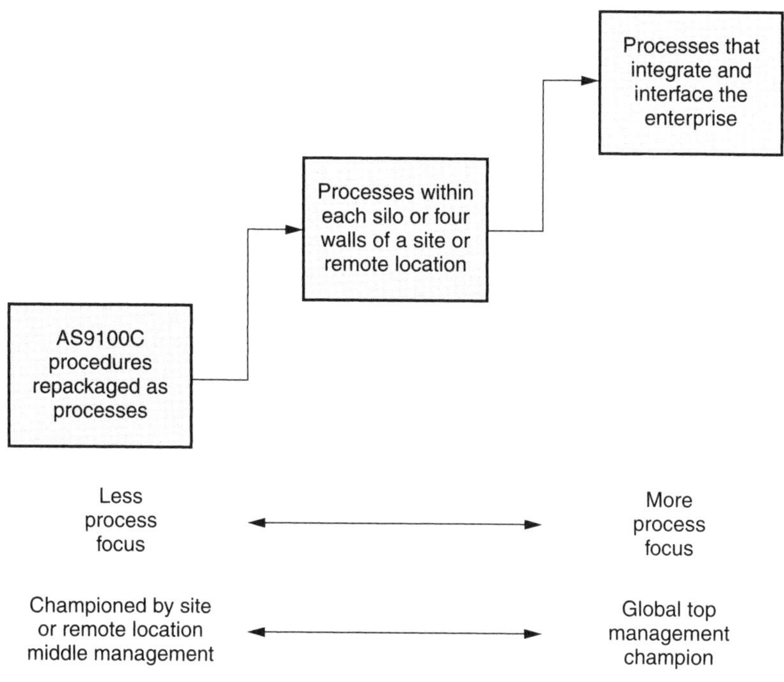

Figure 5.4 Process map showing information/action flows between sites.

Process Map and Process Interfaces

Process interfaces are important, not only between the sites and the supporting functions, but also between processes within the process map. For example, in the process map depicted in Figure 5.1, the marketing and sales process links with the product realization (or business creation) process. What does that link mean in the actual process document? Does the process documentation that describes the sales and marketing process explain what type of information is exchanged between the two processes and define the input and output between the two processes? When one studies the business planning process, is the same information in the sales and marketing process shown as an input into the business planning process? This is what's meant by process interfaces—two processes, which are linked in the process map, that actually work together, as documented, in a real operation.

Measuring and Monitoring Processes

A process-focused organization can explain and show how each process is measured, monitored, and improved. This organization will have not only a process map that includes COPs and has processes that link between the sites and remote locations, but also have evidence of process documentation, measurement, and monitoring, and an improvement plan. Process measurements must gauge the effectiveness and efficiency of a process and track performance.

An organization must establish a process name, definition, measurement, process champion, control method, process goal or objective, and means of tracking progress toward the goal or objective. If an organization isn't making progress, it needs to have an improvement plan.

Relationship between Process Performance and Overall Performance

After the auditor reviews the process map or its equivalent, he or she should study the sequence and interaction between overall performance and process performance. For the sake of discussion, overall performance will be called "results," and the measurement of results will be called "result measurables." The auditor should determine exactly which processes help the organization get results. For example, if there's a problem in delivery, what processes are critical to helping the organization meet customer deliveries? The answer may be processes such as maintenance (which affects uptime), staffing (key jobs should be adequately staffed), and/or supplier management (that is, supplier parts availability). The auditor should be able to study the process data supplied and interview personnel during the readiness review to ascertain suspect processes.

Auditing Each Process

Essential Process Characteristics

AS9101D clauses 4.1.2.4, Process Management, and 4.1.2.5, Process Performance and Effectiveness, identify several factors important to each process, including:

- Process responsibilities assigned and responsible functions assigned
- Process sequencing and interactions established
- Process measurements against requirements and defined measures
- Availability of resources and information

- Training and effectiveness
- Process monitored, measured, and analyzed against planned results (process effectiveness)
- Actions implemented to achieve planned results and to promote continual improvement
- Records maintained

Internal auditors can use the following two tools to audit the process characteristics. The BMS control plan can be used during the stage 1 audit and can be filled out by the auditee before the audit, and the Turtle diagram can be used by the auditor in stage 2 in conjunction with the PEAR.

BMS Control Plan

This tool can be used to determine how processes are managed in an organization. It can be filled out by the auditee during stage 1 and can be used by the auditor to evaluate processes. Figure 5.5 depicts an example of a BMS control plan.

The presence or lack of an item in a column is an indication of the presence or lack of process management. The choice of process measurable needs to be related to the expectation of the process.

*Turtle Diagrams**

An auditor can use Turtle diagrams (Figure 5.6) to evaluate and analyze processes. Turtle diagrams specify a process input and a process output, so they clarify the AS9101D process definition. When evaluating a process, ask what triggers the input and output. The answer is placed in the center of the diagram to form the Turtle's "body." To form the Turtle's "legs," identify the supporting components of the process, that is, who's involved in operating the process, competency levels of the personnel involved, materials and equipment needed to execute the process, methods and procedures needed to operate the process, and, finally, criteria or measurements needed to monitor and improve the process.

Interviewing the auditee and analyzing the process allows an auditor to move away from the checklist and ask more questions about the process itself. The BMS control plan and the Turtle diagram are tools that an

* AS9100 Lead Auditor training offered several process audit tools, including Turtle diagrams, with these Turtle diagrams being the most common and useful process audit tool. Therefore, they will be used in this book for AS9100 internal auditing.

Business Management System Control Plan								
Organization ______					Product description ______			
General manager ______					Issue/rev date ______			
Process activity	**Customer requirement/ expectation**	**Key process/ COP**	**Measurement**	**Responsibility**	**Acceptance criteria**	**Review frequency**	**Control methods**	**Comments/ reaction**
					Q1, Q2, Q3, Q4			
				Examples				
Business fulfillment	On-time delivery	K	% on time in operations	Logistics	94, 95, 96, 96%	4/yr	Monthly management meeting	Corrective action after 3 consecutive
			% on time to customer	Production control	100%	4/yr	Trend chart	C/A is more than 15% off target
Customer complaint	Provide timely response	C	Complaint response	Quality	10 days	4/yr	Production control department	Continue to monitor
Design and development	Meet timing requirement		Time to market	Design/ development	52 weeks	Weekly	Quality department	R, Y, G reaction
Business creation	Innovation		Patents filed	Design	20 per year	Monthly	Design meeting	Continue to monitor

Figure 5.5 BMS control plan example.

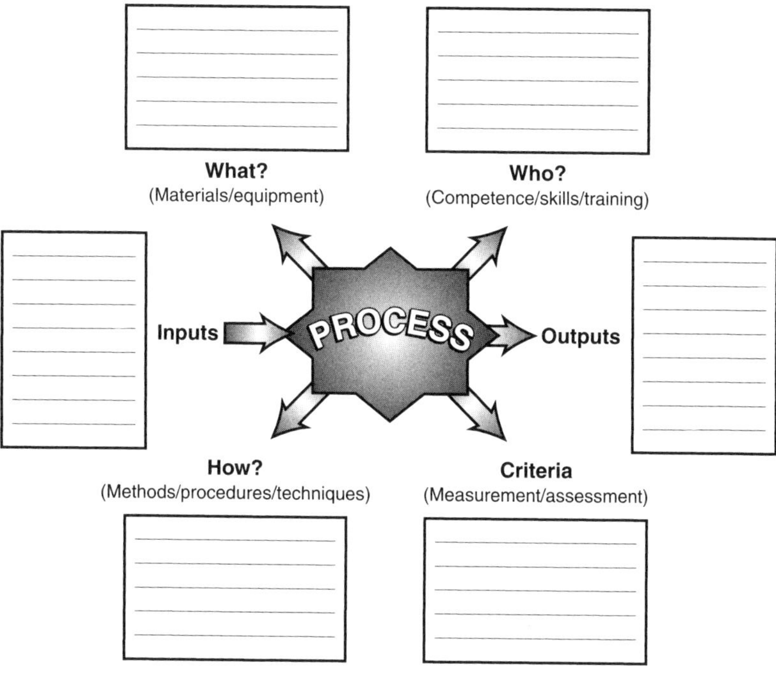

Figure 5.6 Turtle diagram example.

aerospace auditor can use to determine process definition and effectiveness. Neither are specifically required by the AS9100 or AS9101 standards.

Evaluating the Process Approach

Overall, the process map needs to include all the processes that the organization uses to manage and operate the business satisfactorily, such as new product development, management, and support. The key pitfalls here are a clause-oriented process map that really only documents the standard, or a process map that represents the organizational chart. Both of these types of process maps are not representative of the organization. At the end of the day, the process map needs to be reflective of the organization it represents as well as the way the business operates.

The processes also need to both satisfy the requirements of the standard and also be reflective of the organization. Moreover, the process map needs to fulfill the requirements of the essential process characteristics. Using the BMS control plan and Turtle diagram will help determine the process approach. The documentation review of the processes, or level II procedures, will help ensure that the processes "address all the requirements" of AS9100/AS9110/AS9120.

More information on both the process map and processes can be found in Chapter 7, Stage I Audit.

CUSTOMER FOCUS

AS9101D clause 4.1.2.1, Customer Focus, requires customer feedback to be audited as a process. Customer feedback is both an input and an output to the organization; in other words, it encompasses both clauses 5.2, Customer Focus, and 8.2.1, Customer Satisfaction.

It's easy to confuse clause 5.2, Customer Focus, (that is, input) with clause 8.2.1, Customer Satisfaction, (that is, output). It's the organization's interactions with the customer that result in customer satisfaction. In other words, customer needs and expectations are an input, and customer satisfaction is an output. Of course, there's a feedback loop from customer satisfaction back to customer needs and expectations so that the organization can be aware of which customer expectations aren't being met. This loop can be seen in Figure 5.7.

But both customer input and customer output are customer feedbacks that need to be considered by the organization.

Clause 5.2, Customer Focus, requires that an organization study the "needs and expectations" of its customer(s). Accordingly, the auditor should expect to find a process that describes how the organization captures all customer inputs/feedback. Examples of customer feedback given in clause 4.1.2.1 of AS9101D include nonconformity data, corrective action requests, results of satisfaction surveys, complaints regarding product quality, on-time delivery, service provision, and responsiveness to customer and internal requests, including the IAQG Online Aerospace Supplier Information System (OASIS) database. Furthermore, clause 5.2 requires that product

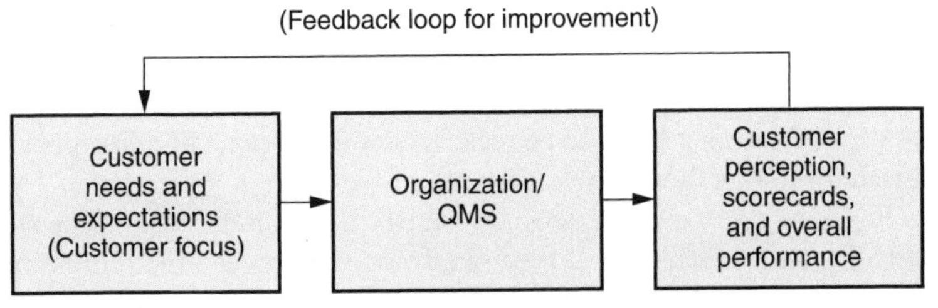

Figure 5.7 Customer focus and customer satisfaction.

conformity and on-time delivery are measured, and action taken if the planned results are not achieved. The process should explain how the organization uses the data gathered in terms of meeting customer requirements. (*Note:* It's not enough for the organization to show evidence of a few interactions with the customer. Documented evidence must be shown that data have been gathered and an analysis completed to indicate that actions are being taken to sufficiently satisfy the customer.)

When analyzing customer focus, the auditor should keep in mind that customers typically suggest some form of product quality, delivery, and cost as important expectations. But there are usually other expectations. What else is the organization learning about customer expectations? Is the organization establishing objectives based on those expectations? Is the analysis of customer needs leading the organization to short- and long-term actions? These actions are encompassed in the business planning, and should typically impact the objectives of the organization.

Auditing Customer Focus

Study Customer Performance

Look at performance expectations through customer scorecards, policy, performance objectives, and past audit results. The performance expectations of AS9101D include nonconformity data; corrective action requests; results of satisfaction surveys; complaints regarding product quality; on-time delivery; service provision; responsiveness to customer and internal requests, including customer satisfaction (perception), product conformance and on-time delivery, customer complaints, and problem solving; and overall performance. All of these issues can create risks to the customer or organization. The audit for AS9100 begins with the auditor analyzing overall customer satisfaction and organizational performance, based on the above criteria. The results of the analysis are linked to suspect or ineffective processes, which are then documented in the aerospace process approach audit plan. The auditor also takes these results into consideration when auditing in accordance with clause 5.2, Customer Focus, and investigates what the organization is doing when performance falls short. The auditor follows that audit trail to determine how the organization sets and deploys objectives and manages its business planning. (See Chapter 6 for more information on audit trails.)

Clause 8.2.1, Customer Satisfaction

Clause 8.2.1 of AS9100C requires that organizations "monitor information relating to the perception of the customer as to whether the organization has met customer requirements." This should occur at least once a year—preferably as a continual process—coinciding with the regular business planning cycle. Also, it's important for the auditor to keep in mind that customer scorecards aren't the same as customer perception. Two notes provided in clause 3.1.4 of ISO 9000:2005 have this to say on the matter:

> Note 1—Customer complaints are a common indicator of low customer satisfaction, but their absence does not necessarily imply high customer satisfaction.
>
> Note 2—Even when customer requirements have been agreed upon with the customer and fulfilled, this does not necessarily ensure high customer satisfaction.

In other words, an organization having no problems with customer metrics or scorecards doesn't necessarily mean their customers are satisfied. Satisfaction can be gauged only by asking customers how satisfied they are.

In the AS9100 series, in both clauses 5.2 and 8.2.1, product conformity and on-time delivery have been added as key customer performance objectives. Of course, AS9101D clause 4.1.2.1, Customer Focus, has added other key areas of customer performance information as noted above. In other words, a system's effectiveness is measured by its ability to satisfy customers and meet organizational performance goals.

Customer notifications are the supplier-rating classifications employed by customers. Generally, these designations are not common in the aerospace industry, but some examples include "needs improvement," "new business hold," or other such designations. Any of these classifications are indications of customer dissatisfaction. Top management should be notified promptly, through management review, of changes to these classifications.

Customer Scorecards

Customer scorecards typically refer to the supplier-rating system of the aerospace industry. Poor customer scorecards and poor performance relative to customer expectations indicate potentially severe customer dissatisfaction. An auditor is required to study performance relative to all customers, especially the key customers of an organization. If they have an online scorecard, the auditor is required to study the most current data and ascertain the supplier score.

Be careful when reviewing these data because each customer has a different rating system. Oftentimes, a customer is happy with only one commodity or set of parts and not another. It's easy for auditors to be misled if they're shown only good data without really understanding the products and/or customer scorecards. It's important that auditors familiarize themselves with all the products supplied to a particular customer and ensure that the data relative to all the products are reviewed on the customer website.

An auditor should make a record of the quality issues, warranty issues, and delivery problems identified in the scorecards, product by product. Each of these issues needs to be tracked down and audited.

Customer Complaints and Problem-Solving Efforts

Any large quantity of product that's been rejected or major quality issue is probably in the scorecard already and should be flagged as a key area to audit. An auditor should look for patterns of the same problem repeating among common parts, and/or many seemingly random returns, which may indicate a lack of process control on the production line for a single product.

The auditor should also study how the organization prioritizes issues, such as whether or not it uses Pareto diagrams. Does the organization analyze data by part family or process family? If it doesn't analyze data meaningfully for problem solving, that could be a nonconformance or an opportunity for improvement. Continual problems also indicate poor problem-solving capabilities or process control issues. Find out how the organization calculates parts-per-million (ppm) external defect levels. Parts-per-million defect levels as high as 30,000 ppm or more are not unusual in the aerospace industry, but the goal of organizations should be to reduce this to 100 ppm or less. Keep in mind that world-class performance, such as in the semiconductor industry, where Omnex has much experience, is in the 10 ppm range. The auditor should analyze the customer complaint database to sample issues and establish an audit trail. Continual problems on one issue or part number—or many different problems for one part, part family, or department—are potential samples for the auditor.

The key to managing customer-related problems and issues is analyzing customer-related data and responding promptly to customer issues.

When assessing an organization's readiness, an auditor should evaluate not only customer-related issues, but also how the organization responds to them.

Identifying Customer Risks

Overall, customer-satisfaction and scorecard issues, and customer complaints about product quality, all constitute risk to the customer. These issues should all be investigated during the on-site audit. Internal audit performance and overall performance issues constitute risks to the customer as well. Internal issues will often translate to customer issues. Identifying suspect processes allows auditors to translate these issues to on-site process investigations during the stage 2 audit. This is described later in this chapter.

Evaluate Internal Audit and Management Review Results

Evaluating internal audit and management results from the previous 12 months allows the auditor to focus on how the organization monitors and understands itself. The auditor reviews the internal audit to ensure that the organization has conducted a complete system audit that includes all its processes as well as all the clauses of the AS9100 series being audited. The organization's internal auditing process must include, at a minimum, a system audit and special process audit. Best-in-class auditing includes manufacturing process and product audits, and layered process audits.

Internal auditors have to be trained to conduct audits. Study the organization's competency requirements for internal auditors.

For best practice, the audit must cover all work shifts and should be based on customer focus and process approach, and include a leadership audit. Audits should be scheduled based on status, importance, and the annual plan. Also, the audits should focus on customer complaints, internal and external performance data, and how the internal audit has considered customer-specific quality management system requirements.

Study the quality of the audit and the nonconformities issued. Is the audit adequate? Does the internal audit include all the issues noticed in the organization thus far? Any nonconformities issued should have three parts: nonconformity, a direct quote of the requirement, and the objective evidence. What's the quality of the nonconformities? Are they clear and concise? (See Chapter 8 for details.)

Check the quality of the nonconformity closeouts. Is there objective evidence to show that the proposed corrective action has been implemented? Also, is there evidence that the system's corrective actions have been implemented? Is there evidence to show that the problem won't be repeated?

System Audits

System audits aren't a series of short audits conducted monthly, but rather a snapshot in time of the overall health and vitality of the QMS. They should be conducted periodically, at a minimum once a year. System audits are done with the same formality as third-party audits and should make use of the same processes and time durations as an initial audit. System audits should cover all the process map processes and all the clauses in the AS9100 series being audited, and they should use the aerospace process approach to audits described in this chapter.

The intent of these audits is to ascertain whether the overall system is "effective." This is the formal audit, which should be conducted similarly to an external audit. During this audit, auditors ensure that the organization is moving toward its goals and objectives, and that customer satisfaction, including product-conformity and on-time delivery metrics, is being addressed.

Process Approach versus Clause or Element Approach

Very simply, auditors conducting previous AS9100 series audits used a clause or element approach. The audit plan included the names of clauses or elements that the auditor would audit. With AS9101D, the audit must follow the aerospace process approach to audits. The audit plan must include processes from the process map. Processes aren't chosen at random, but are prioritized based on risks to the customer (nonconformity data; corrective action requests; results of satisfaction surveys; complaints regarding product quality; OTD; service provision; responsiveness to customer and internal requests, including customer satisfaction (perception), customer complaints, and problem solving; and overall performance).

Overall Performance

An auditor should gauge an organization's overall performance by examining records of management reviews. *Note:* Sometimes auditees conduct a management review only once a year to satisfy the AS9100 series, and they do it only to show the auditor. This type of compliance to the AS9100 series of standards for an important requirement should be duly recognized as a major nonconformity.

The management review should be conducted at suitable intervals to assess overall improvements and to note whether the organization is moving ahead in meeting business objectives and satisfying its customer needs and expectations. It's important for the auditor to note whether the management review is just a presentation of facts or whether it's a meeting that's improvement-oriented and evaluates the need for change to the overall management system, quality policy, and objectives.

At a minimum, the management review or business review must cover these topics:

- Clause 5.6.2, Review Input
  - Results of audits
  - Customer feedback
    - Clause 8.2.1, Customer Satisfaction (customer data and perception shall be monitored to ensure they meet customer needs)

- Process performance and product conformity
- Status of preventive and corrective actions
- Follow-up actions from previous management reviews
- Planned changes that could affect the QMS
- Recommendations for improvement

The auditor shouldn't expect the organization to cover each topic in every business review. However, to move the organization forward, the topics need to be covered according to top management wishes for the frequency of reviews of different topics.

Also important to the management review evaluation is to verify that management review output includes decisions and actions for improving the QMS, processes, product related to customer requirements, and resource issues.

Overall performance should be gauged by examining the weekly and monthly business reviews. Review key indicators of the business, and note indicators of poor performance. Assess the overall quality of the business reviews. Is the company progressing toward its objectives? Also, do the objectives reflect customer needs, expectations, and key concerns?

Measuring Key Indicators and Performance Trends

Best practices for management review means an auditor can expect the organization to use a trend chart that shows the variable being measured on the *y*-axis and the months on the *x*-axis. Omnex recommends the use of trend charts, Pareto diagrams, summary of actions taken, and Paynter charts (see Figure 5.9).

There is also a relationship between results and process performance. Performance trends are directly related to the performance of the processes, as shown in Figures 5.8 and 5.10.

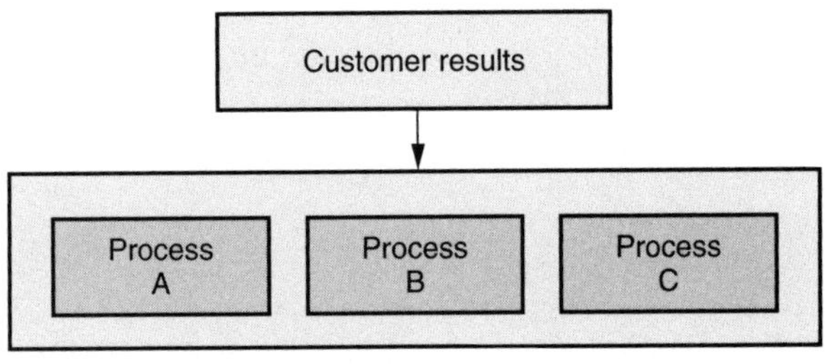

Figure 5.8 Customer results.

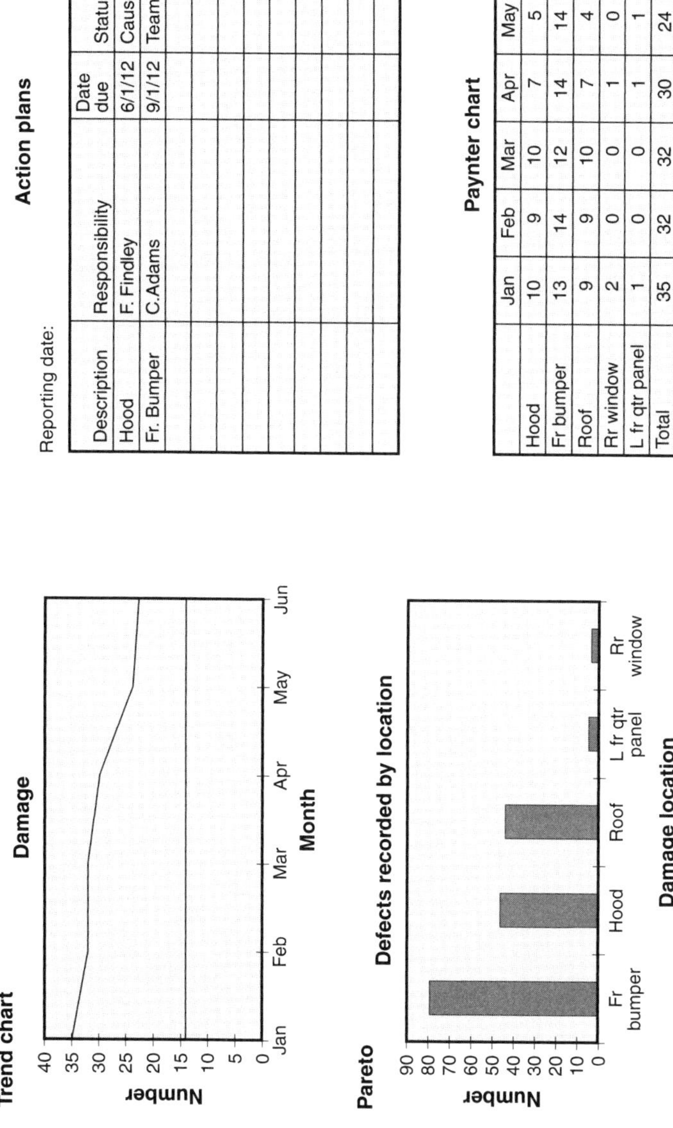

Action plans

Reporting date:

Description	Responsibility	Date due	Status/inhibitors
Hood	F. Findley	6/1/12	Causes identified
Fr. Bumper	C.Adams	9/1/12	Team formed

Paynter chart

	Jan	Feb	Mar	Apr	May	Jun	Total
Hood	10	9	10	7	5	5	46
Fr bumper	13	14	12	14	14	12	79
Roof	9	9	10	7	4	5	44
Rr window	2	0	0	1	0	0	3
L fr qtr panel	1	0	0	1	1	1	4
Total	35	32	32	30	24	23	176

Figure 5.9 Recommended charts for key indicators and performance trends.

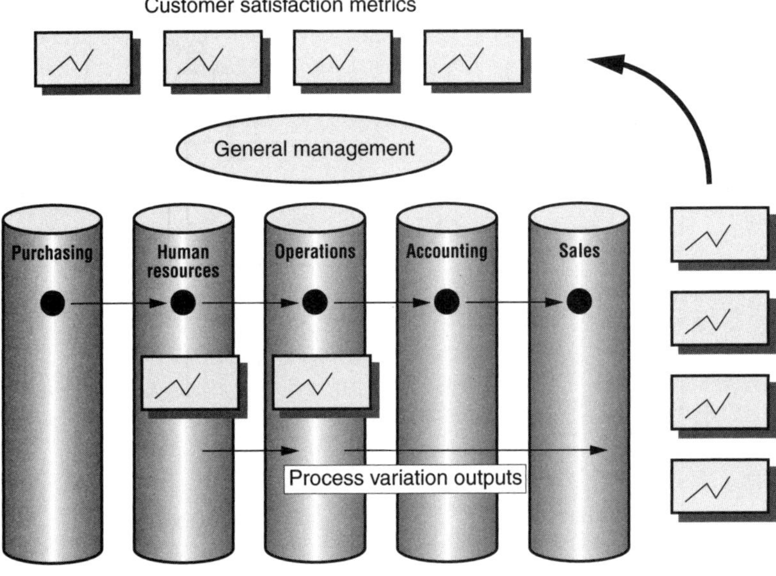

Figure 5.10 Functions versus processes versus results.

Identify Suspect Processes

When identifying processes suspected of poor performance based on customer and performance data analysis, an auditor looks at how the organization links results to process performance. In particular, he or she should gather information regarding the following concerns:

- Overall satisfaction
- Customer scorecards
- Customer complaints
- Overall performance

The auditor should write down the issues that present a risk to the customer and organization, and then try to group these risks into common categories. An example of grouping common risk categories is shown in Figure 5.11.

Suspect processes are those that indicate poor performance (that is, present risks to the customer and organization) based on customer and performance data analysis. Once the performance problems are identified, the auditor should identify what the suspect processes might be.

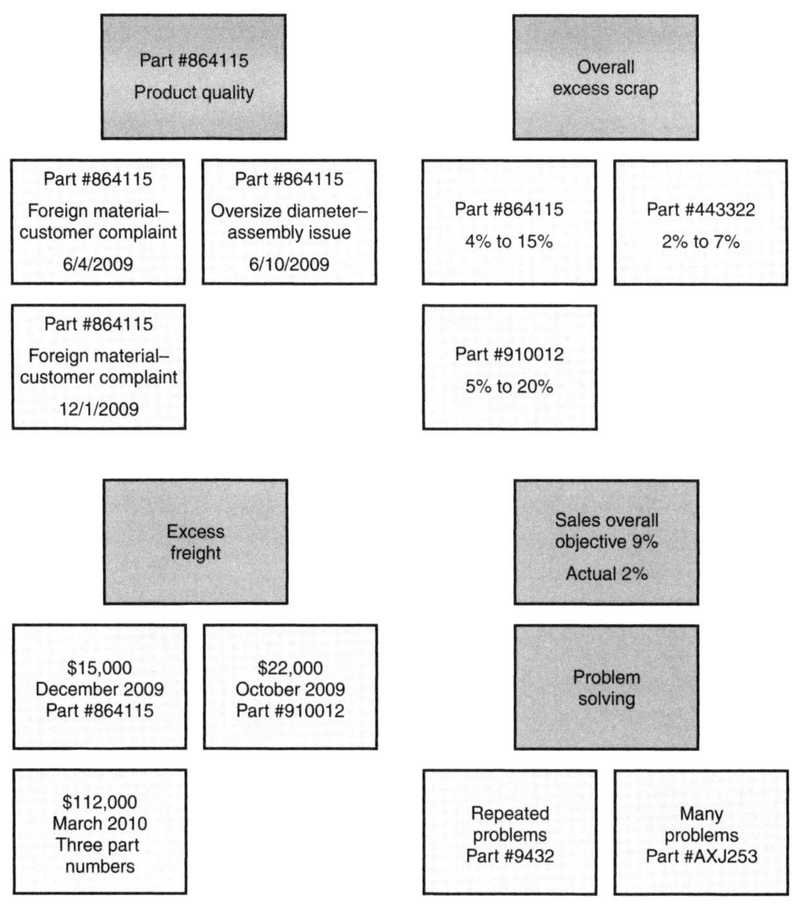

Figure 5.11 Grouping risks into common categories—example.

Leadership

The third piece to aerospace auditing is the interview with top management. This is conducted during the stage 2 on-site audit. It is sometimes difficult for internal auditors to interview their top management. For this purpose, Omnex often recommends that there are first-line reports to top management on the audit team.

Before the interview, the management representative will want to coach top management and let them know that they will be interviewed. This will make it easier for the internal auditor to conduct the interview.

Meeting with Top Management

The following clauses of AS9100 relate to top management:

- *5.1, Management Commitment.* Asks top management to personally accomplish certain tasks and to direct and enable others.
- *5.2, Customer Focus.* Requires top management to understand customer expectations and help meet them.
- *5.3, Quality Policy.* Top management needs to help ensure that a quality policy is established and needs to set objectives that align with that policy.
- *5.4, Planning.* Top management needs to set measurable objectives that are deployed within the organization, as well as create a plan to meet those objectives.
- *5.6, Management Review.* Top management participates in reviewing data required by AS9100 to help the organization improve and change.
- *8.2.1, Customer Satisfaction.* Requires top management to review customer satisfaction and make decisions or take actions for improvement.

Note: Certain top management activities can be delegated; whenever the word "ensure" is used (for example, in clauses 5.2, Customer Focus, 5.3, Quality Policy, and 5.4, Planning), these items can be delegated.

Interview with Top Management

The interview(s) with top management evaluates:

- The establishment and continued relevance of the organization's quality policy and objectives
- The establishment of performance measures aligned to quality objectives
- Quality management system development, implementation, and continual improvement
- Top management commitment
- Quality management system performance and effectiveness
- Performance to customer expectations (for example, supplier rating, scorecard, audit results)

- Actions taken to address areas that are not meeting customer performance expectations

After the interview, determine how well the top manager answered the relevant questions asked regarding the QMS and its performance, and also about customers and the organization's success in achieving customer satisfaction. The interview and top management's responses will gauge top management commitment and involvement in the quality management system.

SUMMARY

Process focus, customer focus, and leadership audits are key to the aerospace approach to audits. The process focus ensures that the organizational processes are represented in the QMS and that they satisfy the AS9100 series requirements. Secondly, it ensures that the organization is not only process focused, but is also focused on the customer, and lastly, that top management has a good understanding of the QMS, its processes, and the overall customer issues. These are the three important factors in the overall success of an organization, and they form the crux of the aerospace approach. Internal auditors should read and understand this section before proceeding to Chapter 7 on the stage 1 audit.

6

Understanding Audit Trails

AUDIT TRAILS AND CUSTOMER ORIENTED PROCESSES

This chapter introduces three basic audit trails that describe key links for the major types of processes found in most organizations. Because the actual audit trails will vary from one organization to another, using tools such as audit trail flowcharts will aid the auditor in confirming the actual processes and links to audit.

Audit trails are techniques that minimize the number of times an auditor visits one area. They also allow for a detailed study of links because the processes connected by samples taken in one area may lead to a second or third area of investigation. For example, during the business planning and management review, an auditor would find customer needs and expectations in the process that gauges customer focus. The auditor would subsequently move on to the quality policy in relation to that customer focus. This could then lead the auditor to study the business objectives as they relate to needs, expectations, and quality policy. Finally, the management review would be studied to ensure that the company is reviewing and fulfilling the objectives of the business, and that it's ultimately satisfying the needs and expectations expressed by its customers.

The three audit trails to be utilized when auditing are:

- *Business planning and management (BPM) review.* This audit trail is used when auditing strategic planning, business planning, policy deployment, objective setting, customer expectations, management review, or operations review.

- *New product development (NPD).* This audit trail is used when auditing processes related to new product development. This process begins with the bid/quote and ends with the product approval.
- *Provision.* This audit trail explains the links between clauses during auditing of manufacturing or operations.

Readers will note the similarities between the process map blocks discussed in the previous chapter and displayed in Figure 6.1 and the audit trails discussed in this chapter.

Finally, an auditing technique that uses the Turtle diagram and PEAR form, called *process monitoring and improvement,* will be introduced toward the end of this chapter. This technique will be available to audit all processes, selected processes, or just processes in clause 7.0, Product Realization, as suggested by AS9101D.

HOW TO USE AUDIT TRAILS WHEN AUDITING PROCESSES

The auditor must develop an audit plan by using processes instead of clauses. Organizations using clause-based audit plans will not really be process oriented. The auditor should study the organization's process map or interview the auditee during the readiness review to understand what processes satisfy specific clauses of the audit trail. He or she should then insert the processes into their respective audit trails. In other words, the auditor should use the company's business processes during the audit for guidance, and not use the AS9100/AS9110/AS9120 requirements for guidance.

The business planning and management review (or management oriented process [MOP]), new product development (which falls under customer oriented process [COP]), and the provision audit trail (also a COP) are three important audit trails. Every system audit should include these three mission-critical audit trails.

The organization's two product realization processes are new product development and the provision audit trail. The new product development process typically cuts across several support locations, including sales, program planning, design engineering, product engineering, and product and process validation. Meanwhile, the provision audit trail takes place at the manfacturing site.

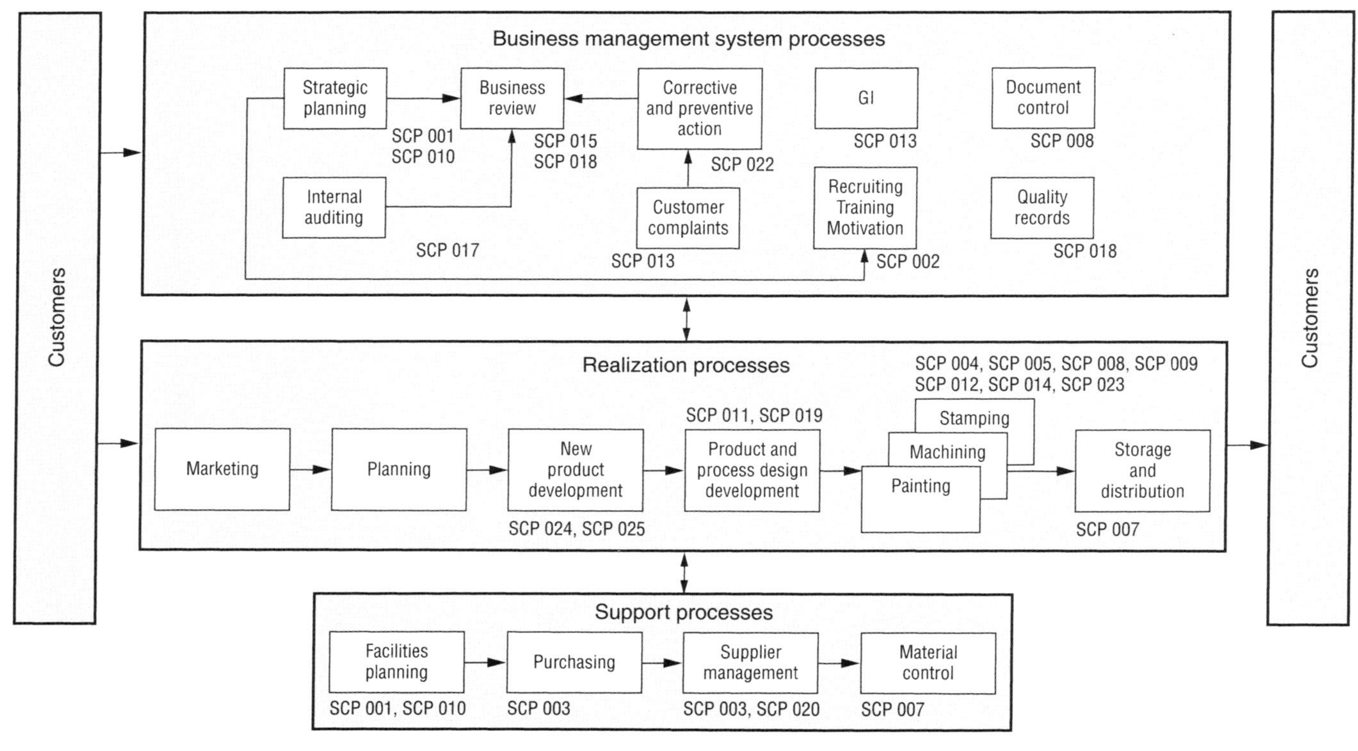

Figure 6.1 Business management processes.

BUSINESS PLANNING AND MANAGEMENT (BPM) REVIEW AUDIT TRAIL

The business planning and management review audit trail is one of the most important. If this audit trail is effective, then there's a good chance the other clauses of AS9100/9110/9120 will also be well implemented in the organization. Management commitment, along with the business planning process, often guarantees an organization's success or failure.

In the flowchart in Figure 6.2, Customer Focus (clause 5.2) is the starting point for this audit trail.

The phases of the business planning and management review audit trail can be broken down as follows:

- The auditor starts the audit with top management, learning its strategic planning process. The auditor samples customer expectations and requirements to evaluate the audit trail.
- The auditor uses expectations and requirements to check their alignment with the quality policy (clause 5.3), and then looks at both as they relate to the quality objectives chosen (clause 5.4.1).

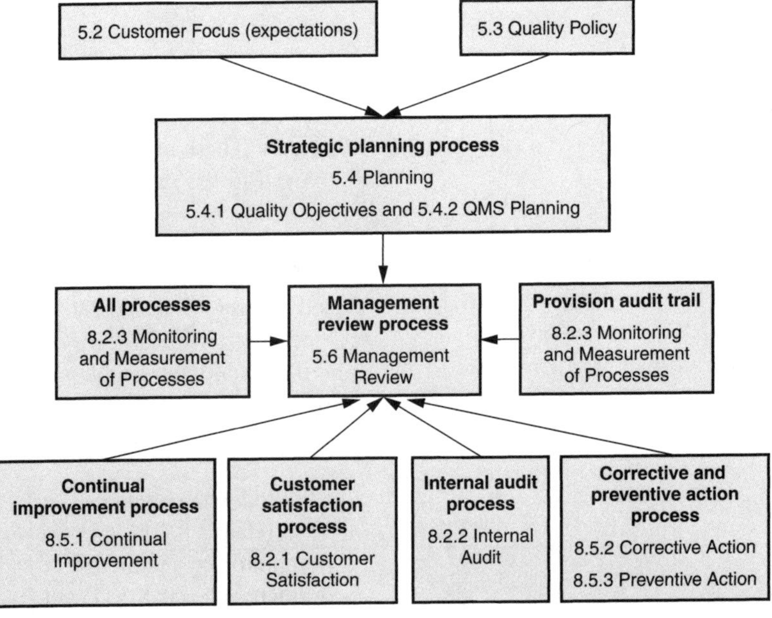

Figure 6.2 Business planning and management review audit trail.

- The quality objectives drive the quality management system (QMS) planning (clause 5.4.2), which results in a continual improvement plan and QMS changes.
- The plan (clause 5.4.2) also needs to include new or changed processes under general requirements (clause 4.1).
- The auditor then examines the internal audit process (clause 8.2.2), the customer satisfaction process (clause 8.2.1), continual improvement (clause 8.5.1), and corrective and preventive action processes (clauses 8.5.2 and 8.5.3).
- The results of these processes are reviewed during the management review. If the process monitoring audit trail and/or operations audit trail are assigned to the auditor, then they should be completed before the management review is audited.
- Finally, the management review process is audited to ensure that the QMS is being managed effectively and that performance improvements are taking place. The auditor can compare the organization's set objectives against its actual performance. The auditor can also gauge whether the organization is truly satisfying customer needs and expectations.

Links and Samples

It's important to recognize that, in order for an organization to be considered customer focused, good practice requires key customer requirements (for example, on-time delivery) to have objectives. These must be included in the organization's business plan and deployed to the remainder of the organization.

These objectives need to have continual improvement goals and appopriate resources. The objectives should also be linked to customer expectations and be available in a defined time period. Therefore, it isn't enough to sample some customer requirements and objectives. The auditor must start with key customer requirements, follow the trail, and then sample related documents. It's essential that the audit trail (see Figure 6.3) is followed thoroughly and completely. The business plan objectives are key pieces of evidence for this audit trail.

The second set of links is reviewed in the following sequence: internal audits (clause 8.2.2), customer satisfaction (clause 8.2.1), monitoring and measurement of processes (clause 8.2.3), monitoring and measurement of product (clause 8.2.4), continual improvement (clause 8.5.1), corrective action (clause 8.5.2), preventive action (clause 8.5.3), and management review (clause 5.6).

Customer Expectation Sampling Sheet

Customer expectations* (5.2)	Objectives (5.4.1)	Deployed objectives or departments (5.4.1)	Related processes (4.1 or 7.0)	Strategy to meet objectives (5.4.2)	Evidence of objectives being met (5.6 or 8.5.1)
Delivery*	100 percent on time	Purchasing: Supplier shipment 100 percent on time Scheduling: No change in schedule two weeks out ——— Quality: N/A	• Supplier development process • Supplier rating • Scheduling process	• Project to increase uptime from 84 percent to 95 percent • Project to increase supplier short shipments	• No improvement seen. Project initiated two months ago. • Supplier short shipment shows improvement of 25 percent
Reduce product lead time*	Reduce lead time from twenty-four months to sixteen for major product launch	Purchasing: Involve suppliers up front Scheduling: Reduce supplier submission misses to zero ——— Quality: none	• Project planning • Design strategy process	• Project to reuse technology • Project to increase simultaneous engineering	• Reuse technology in process • Improvement of 20 percent is evident ——— • Simultaneous engineering project implementation in process • No requirement evident

* Are you sampling product quality, delivery, technology, lead-time, reliability, and cost?
* Note: Which dimensions are important to the customer?

Figure 6.3 Customer Expectation Sampling Sheet example.

The processes measured in clause 8.2.3 are the same as those identified by following the customer requirements.

The auditor should be able to study the process and ask whether the organization measured the right criteria in the process. For manufacturing processes, the auditor must sample specific process information (that is, process studies and capability).

Samples of five to 10 key customer requirements should be selected to audit this trail appropriately. The auditor should use the Customer Expectation Sampling Sheet in Figure 6.3 to follow the trail from key customer expectations to objectives, deployed objectives, related processes, a plan for meeting the objectives, an improvement plan that fulfills the objectives, and, finally, to evidence that the continual improvement is occurring. On this audit trail, the auditor starts with customers and ends with improvements made to meet the customers' needs.

Examples of processes include:

- Customer feedback (includes customer expectations, contract review, and customer satisfaction)
- Market research
- Business and/or strategic planning
- Management review
- Operations review

The clauses for BPM audit trails are those associated with top management. Top management's responsibilities are defined in clause 5.1, Management Commitment. There are only two responsibilities that top management can delegate according to clause 5.1: ensuring that quality objectives are established and that resources are made available. All other tasks listed in clause 5.1 must have top management's direct involvement.

NEW PRODUCT DEVELOPMENT (NPD) AUDIT TRAIL

The success of the new product development process is vital to the future of the organization being audited. This process is ineffective in many organizations, leading to problems with timing, product quality at launch, and designs that don't meet customer expectations.

The audit trail, as seen in Figure 6.4, starts by identifying customer-related processes or sales processes (clause 7.2), then proceeds to program planning, including project management, design and development of the product (clause 7.3), process or technology development, and, finally, to the production part approval process or first article inspection (FAI).

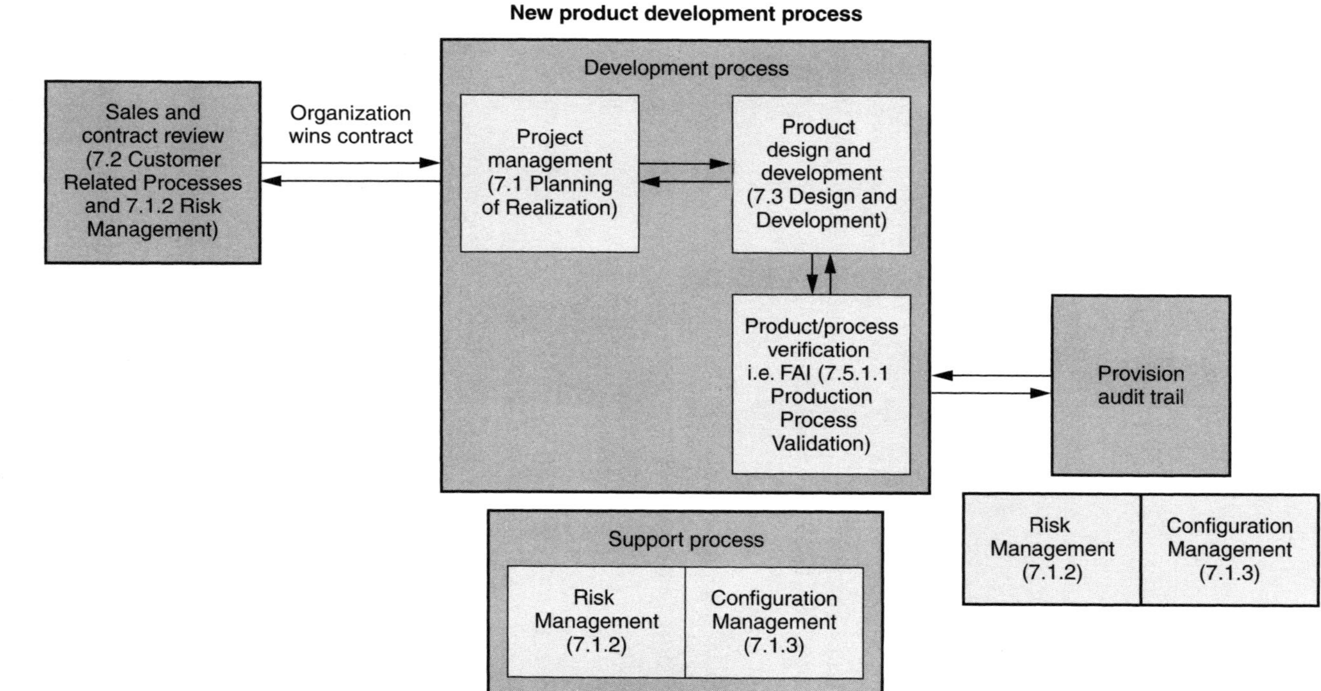

Figure 6.4 New product realization audit trail.

Built in throughout NPD is the concept of risk management, which starts with the contract during sales and finally finishes with the key characteristics controlled in the manufacturing processes either at the organization or their suppliers.

After the FAI, the audit moves to the operations, process monitoring, and improvement audit trails.

Links and Samples

The set of links in this audit trail are those from contract review (clause 7.2) through to the product approval process. To audit this trail, the auditor first selects product samples and tests customer-related processes or sales processes (clause 7.2) for the contract review. This will verify that the organization has properly identified and reviewed the product requirements prior to agreeing to supply the product. Contract review takes place in the sales and marketing function. To ensure that there's a trail to product planning and then to the plant, the auditor will need to identify the products that have recently been through the FAI process and/or are being produced in the plant.

Next, the auditor examines how the organization has planned the processes required to produce the product identified in product realization planning (clause 7.1). This audit of process planning takes place where the organization conducts its project or program management. During this phase, or early in the audit trail, the auditor needs to document the goals and objectives of the product launch. What were the timing, quality goals, and budget at the start of the program? Those in the organization who were involved with program or project management should be able to identify where they started and what the results were. If they can't, that in itself is a finding in the audit.

For a product that was launched a few months to a year before, the organization should provide the auditor with the quality history, or scrap, rework, and customer rejects data. It's important that the auditor conducts this performance evaluation and subsequently prioritizes the audit areas. He or she then tracks the product sampled through design and development (clause 7.3) to test that each step of the new product and process development processes has been followed. Finally, the FAI is reviewed for content and effectiveness.

It's always good to sample a variety of products, including those that have been running for one year, were recently introduced, and are still in the pipeline. The auditor should try to audit each of these, and many others chosen randomly, when auditing customer-related processes (clause 7.2). In this way, contracts that were won and lost will be audited. The auditor can study how customer requirements for a particular product were gathered by the organization, as required in clause 7.2.

The audit scope and product definition are important on this audit trail because a product could include hardware, software, service, and processed material. The auditor should carefully define the scope ahead of time.

Examples of processes include:

- New product development
- Risk management
- Bid and/or quote
- Order entry
- Product design and development
- FAI, or product and process validation

PROVISION AUDIT TRAIL

The provision audit trail covers all the activities necessary to produce, manufacture, and deliver products and services to the customer, as illustrated in Figure 6.5. It includes activities for operating the realization processes, identifying and tracing products and services, handling customer property, preserving product, measuring product and service output, and handling nonconforming product.

The auditor must carefully study the different processes in the organization for production and service provision (clause 7.5) and sample them appropriately. For example, in a manufacturing plant that has heat-treating, machining, paint, and assembly, each of these processes should be sampled.

This audit trail starts with production and service provision (clause 7.5), then tracks through all the other processes in operations. The auditor studies how processes that produce and deliver the product are controlled, identified, measured, and monitored. The customer requirements and characteristics are important during this step of the audit trail. Nonconforming products are segregated and evaluated for disposition. Finally, the performance of the process is analyzed, and projects are chosen for improvement.

Examples of typical processes are:

- Assembly
- Fabrication
- Scheduling

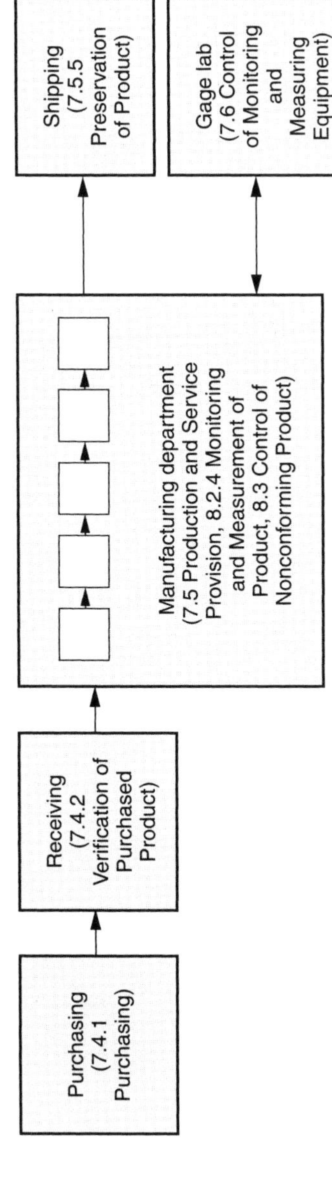

Figure 6.5 Provision audit trail.

- Shipping
- Packaging
- Stamping
- Machining

Links and Samples

The common elements between each of the phases of this process are the products being sampled. The products could be the same ones that were sampled for new product development. The sampling should be conducted with the following considerations:

- Do the sampled products involve all the major processes?
- Do the sampled products represent the major product families?

Select some products that represent a small percentage of sales revenue and some that represent major product lines. This will also determine whether the organization has applied its operational controls across the board. If the organization is required to provide any post-delivery services, that process will need to be sampled as well.

The measurement and monitoring equipment specified in the control plan must be controlled according to clause 7.6, Control of Monitoring and Measuring Equipment. This equipment measures the characteristics in the monitoring and measuring of product (clause 8.2.4). The auditor should document several pieces of monitoring and measuring equipment when auditing production and service provision (clause 7.5), and then follow up in control of monitoring and measuring equipment (clause 7.6) for calibration and measurement systems analysis (MSA) activities.

Clauses 7.6 and 8.2.4 are linked because the former describes the characteristics and the latter measures them. Also, the control plan describes the process for control, and clause 8.2.4 is the mechanism of that control. Another link is between clause 8.2.4, which could measure received products, and clause 7.4, Purchasing, which controls supplier processes.

Finally, after product characteristics are measured in clause 8.2.4, conforming products go to the next stage, and flawed ones are segregated as nonconforming products. These must follow a process described in clause 8.3, Control of Nonconforming Product.

The same product sampled in monitoring and measurement of product (clause 8.2.4) and control of nonconforming product (clause 8.3) is analyzed with other organizational data in the analysis of data (clause 8.4) and improved in continual improvement (clause 8.5.1).

PRODUCTION AND SERVICE PROVISION AUDIT TRAILS

The intent of this audit trail is to verify that an organization has effective control over its production and service operations. To do so, the organization has to set up a number of activities and mechanisms on the production floor. To audit these items, the auditor should proceed as follows:

- Ask for plant layout and a production schedule.
- Sample departments (that is, part numbers) to audit. As a representative sample, include final inspection and packaging.
- Ask for a process flow and control plan/inspection plan.
- Sample processes to audit.

The following assessments should be made for each process sampled in the process flow:

- Study the operator work instructions against the requirements outlined in the control and/or quality plan.
- Ask the operator how he or she operates and sets up the machine.
- Study responses against work instructions.
- Ask how the organization inspects its product.
- Check the inspection checklist or equivalent against the control plan.
- How are both special and regular characteristics inspected and/or controlled?
- Is the frequency and method in the control plan followed (for example, with a check sheet)?
- Ask about gauges mastered and write down gauge numbers.
- Check the gauge calibration sticker or its equivalent for a calibration date. Has it expired?
- Do they have any key characteristics?
- Check statistical process control charts, if any.
- Assess stability and capability.
- Ask questions about out-of-control conditions.

- Check the tagging procedure and/or work instructions. Check visually throughout the department.
- What does the operator do about nonconforming parts? Check this response against a nonconforming product procedure.
- Check inspection status, as appropriate.
- Ask about preventive maintenance as related to the operator.
- Write down operator names, operation numbers, gauge numbers, and document numbers for the audit trail.
- At the end of the interview, ask to see the records. Sample weeks of data for each operation for conformity. (*Note:* This is key to the provision audit trail. Performance to the process controls or inspection plan can only be guaged by evaluating manufacturing records.)
- Ask about the quality policy.
- Ask about process changes.
- Look at the overall cleanliness of the plant and the maintainence process. (*Note:* packaging and shipping areas should be examined as well.)

AUDIT TRAILS AND AUDIT PLANNING

Following the audit trail approach outlined in this chapter should allow the auditor to audit most of the processes defined in a typical AS9100/AS9110/AS9120 process map. In doing so, the auditor will audit not only for conformity, but also for performance as judged by the customer. The auditor won't be auditing processes in isolation, but as they're linked to one another in the normal organizational sense. This approach leads to a true system audit.

The audit trails provide links and interfaces required by AS9100/AS9110/AS9120. However, auditors would be remiss to include these sample audit trails verbatim in their audit plan documents. In fact, such an audit plan could be considered clause-driven, and in a third-party audit situation might result in a major nonconformity.

So what's the best way to use this chapter on audit trails? The auditor should start by analyzing the organization's process maps per the three main audit trails provided—BPM, NPD, and provision. The processes that correspond to the audit trails should be used in the audit plan. The auditor should use the samples to study the integrity of the links and the performance of the system.

The links between the clauses and sampling technique provided in this chapter should be carefully studied. Take a sample process map and then try to understand what processes in the organization correspond to the audit trails provided. Doing this will help uncover weaknesses in the process map.

PROCESS MONITORING AND IMPROVEMENT

Process monitoring and improvement is unlike the audit trails discussed in this chapter. It focuses on how processes such as the BPM, NPD, and provision audit trails are audited. Process monitoring and improvement is used when auditing the processes in a process map. It focuses on overall process control and effectiveness and follows the aerospace process approach to audits.

Process monitoring and improvement is the method used for selected processes, including the three previously discussed. The forms and method provided here will hold the auditor in good stead when auditing processes using the process approach.

Conducting a Process Approach Audit

The process approach is meant to add value. It should start with a performance analysis of customer data that identifies areas of weakness noted during the readiness review. It should end with the identification of variations (that is, nonconformities) in the process that, if eliminated, will lead to process improvement.

Identify process weaknesses and probe them during the audit. Start by asking questions on performance as it relates to the overall weakness in customer scorecards, rejections, or other key performance failures observed. Relevant questions include:

- What are the indicators, objectives, and actual performance?
- How is performance being improved?
- How was the process planned?
- Does the process design allow performance objectives to be met?

- Does the process design meet AS9100/AS9110/AS9120 requirements?
- Does the process design take into account previous performance results?

An example of a process map can be seen in Figure 6.6.

Conduct the process analysis, using a Turtle diagram as necessary (see Figure 6.7). Follow the process using the documented process flow or procedure. Is the process carried out as designed? Are the methods applied? Sample the process, as applicable, where the work is done (for example, engineering, shop floor, or workstation).

After the process audit has been completed, the auditor should be able to determine if the process is sufficiently defined, if it's working satisfactorily, if its performance relates to overall performance, and if the links are well established. The auditor should sample the process to ensure that it's working according to the process definition.

Links and Samples

The links in the process audit relate to organizational performance. As suggested by AS9101D, the auditor is required to prioritize the audit based on the organization's performance. Study customer scorecards, customer rejects, and failures from previous audits. Try to relate these failures to the processes. The process map comes in handy when identifying which processes to sample. This phase is conducted during the stage 1 audit. (For more about this review, see Chapter 7.)

Next, the auditor should conduct a process analysis using a Turtle diagram, such as the one depicted in Figure 6.7, to identify sources of potential variation in the process. The key to an effective process analysis is to determine the actual inputs, outputs, and expectations for this process. In the example already covered in this chapter and shown in Figure 6.8,* the inputs to the business reviews are the agenda items that will be covered in the meeting. The output is the action plan. Internal customer expectations for the outputs are fully attended meetings and actions. The expectations constitute the criteria to determine effectiveness of the process.

Once this is done, the auditor should determine the methods, procedures, material, equipment, competence, skills, and training required for the process to be effective. The process analysis can be conducted prior to, or in conjunction with, the audit. The process analysis will show weakness in the process definition.

---

* Omnex has integrated the Turtle diagram into a combined Process Audit Worksheet/PEAR.

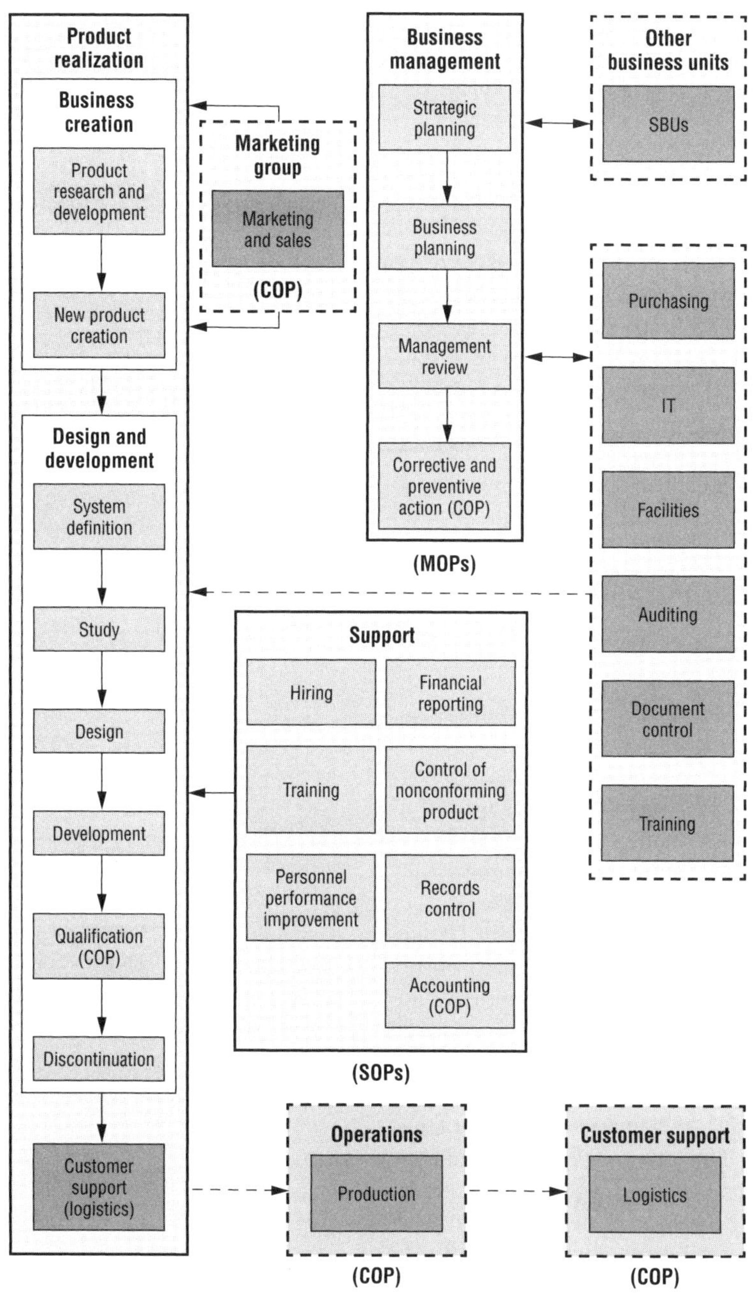

Figure 6.6 Process map example.

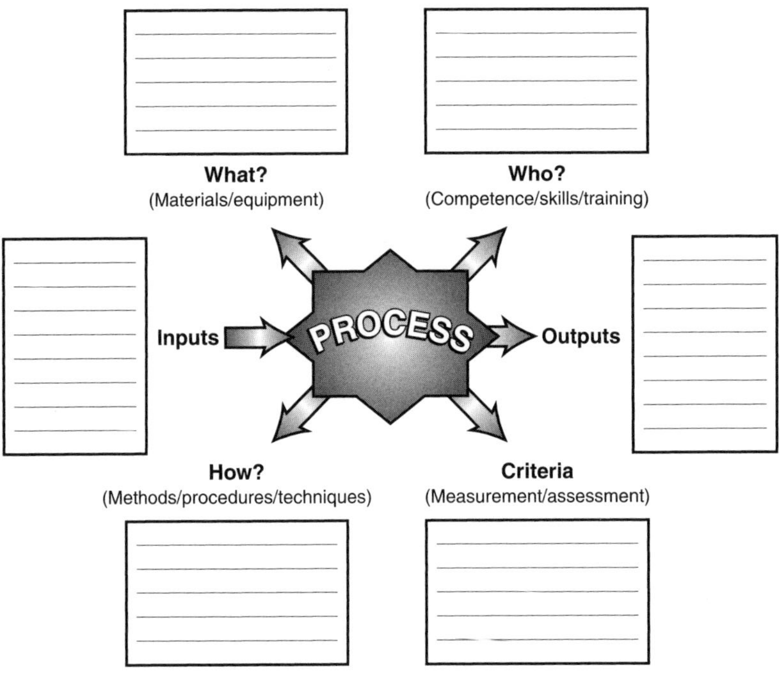

Figure 6.7 Turtle diagram example.

Although the auditor has initiated this analysis during the readiness review, he or she should interview the auditee (that is, process worker) to determine what's actually happening in the process.

If the process analysis for management review shows a key problem uncovered by the performance analysis (for example, lack of actions), then this is an area the auditor will explore further when conducting the audit. Prior to the audit, questions based on the lack of performance should be documented in a process audit worksheet such as the one in Figure 6.8. During the audit, these questions should become the focus of the process approach to audits. In other words, the AS9100/AS9110/AS9120 audit should focus on risks to the customer and organizational performance.

COMPLETING THE PROCESS AUDIT WORKSHEET/PEAR

A Process Audit Worksheet/PEAR, such as the one depicted in Figure 6.8, should be used to record the objective evidence identified during the

Process Audit Worksheet/PEAR

Company name: Ⓐ | **Location:** Ⓑ	**Audit type:** Ⓒ | **Standard:**
Process characteristics: Ⓓ ❑ Has responsibilities/process owner assigned? ❑ Considers customer specifics ❑ Is documented? ❑ Is monitored?	**Classification:** ❑ Needs further research ❑ Nonconformities ❑ Opportunities for improvements
Auditor: Ⓔ	**Process/area:** Ⓕ
Organization's method for determing process effectiveness (for example, key indicators or measurements): Ⓖ	**Applicable clauses (AS9100/9110/9120):** Ⓗ
Auditees/interviewees: Ⓘ **Audit observations and comments supporting process effectiveness determination:**	**Process interfaces (process linkages) and process details:** Ⓙ **Process activities:**
Statement of effectiveness:	1. ❑ Not implemented; planned results are not achieved. Ⓚ 2. ❑ Implemented; planned results are not achieved and appropriate actions not taken. 3. ❑ Implemented; planned results are not achieved, but appropriate actions are being taken. 4. ❑ Implemented; planned results are achieved.

Interactions

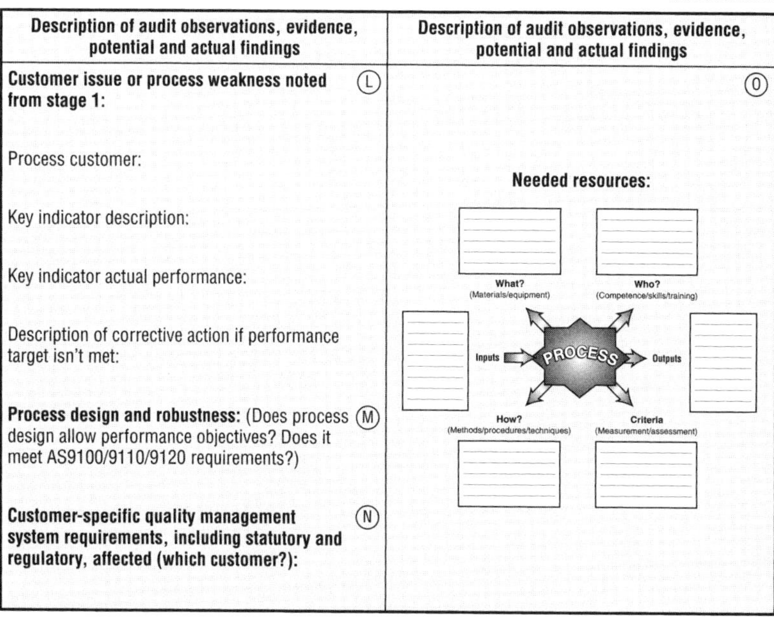

Description of audit observations, evidence, potential and actual findings	**Description of audit observations, evidence, potential and actual findings**
Customer issue or process weakness noted from stage 1: Ⓛ Process customer: Key indicator description: Key indicator actual performance: Description of corrective action if performance target isn't met: **Process design and robustness:** (Does process design allow performance objectives? Does it meet AS9100/9110/9120 requirements?) Ⓜ **Customer-specific quality management system requirements, including statutory and regulatory, affected (which customer?):** Ⓝ	Ⓞ **Needed resources:**

Figure 6.8 Process Audit Worksheet/PEAR example.

process approach audit. One of the key areas of the worksheet is verifying the presence of all process-essential characteristics identified in the previous chapter. The process characteristics are placed on the header of the Process Audit Worksheet.

It's a good idea to complete as much of the Process Audit Worksheet/PEAR as possible prior to the audit for the processes selected. This includes figuring out which questions should be asked. When a suspect or poorly performing process has been identified, it should become the auditor's primary focus.

When completing the Process Audit Worksheet/PEAR be sure to:

- Complete the header with company name and location. (A, B)
- Identify the type of audit—for example, system, process, product, or supplier. (C)
- Identify the auditor by name. (E)
- Identify the process and area being audited. Since it is a prioritized audit, processes chosen typically affect customer satisfaction, customer complaints, or organizational performance issues. (F)
- Identify which process measurements relating to customer concerns (that is, expectations) should be targeted. Also, add the process measurement that they're presently tracking. (G)
- Determine which clauses of AS9100/AS9110/AS9120 the process satisfies. For example, a customer complaint process could have at least two related clauses—8.5.2 and 7.2.3. (H)
- Identify the interfaces or links of the process with other processes. Does the process documentation show the links, and does the client understand them? You should also summarize the process details. (J)
- Determine which customer and customer-specific quality management system requirements affect the process. Also identify all statutory and regulatory requirements. (N)
- Prepare the process-related questions to be asked. (L)
  - First, identify all the process concerns as they relate to performance issues for the customer and/or overall organizational performance. Does the process explain why there's no overall performance recorded in a customer-related issue, scorecard, or satisfaction survey? If not, identify potential questions relating to supporting processes.

- Second, identify the general process-related questions and audit using the Turtle diagram (O). Then determine if the process is performing satisfactorily and whether it's measured, monitored, and improved as required.
- Third, identify potential questions that haven't been asked relating to conformity with AS9100/AS9110/AS9120 requirements. This can be documented in the AS9100C Checklist and Objective Evidence Record (OER) (see Appendix C).

Not all areas of a Process Audit Worksheet/PEAR can be assessed in the stage 1 audit:

- Process characteristics and classifications (D) are determined by the interview in the stage 2 audit.
- The organization's method for process effectiveness (G) is completed during the interview in stage 2. In other words, how is effectiveness measured and maintained? Typically, the process is measured by process indicators or other measurements. See Chapter 2, Figure 2.9 for the BMS control plan.
- Auditor observations and comments supporting process effectiveness determination (I) refer to objective evidence supporting the auditor's conclusion for the statement of effectiveness (K).
- Statement of effectiveness (K) ranges from "not implemented; planned results are not achieved" to "implemented; planned results are achieved."
- Process weaknesses noted for stage 1 (L), although partially completed during stage 2, still needs additional information from the auditor.
- Process design and robustness (M)—if the process repeatedly fails, the auditor needs to ask whether the process has been satisfactorily designed and whether it will need a redesign in order to meet the process objectives.
- The Turtle diagram (O) can be used during the interview to audit the process.

7

Stage 1 Audit

This chapter doesn't take the place of internal or lead auditor training. Instead, it's intended to augment information an auditor would already get in an auditing course. It discusses the aerospace process approach to conducting an AS9100/AS9110/AS9120 audit and focuses on issues not found in a general auditing course based on ISO 9001:2008. What's so special about the aerospace audit for internal and third-party auditors? How do the AS9101D requirements change or augment the existing auditing process normally followed in ISO 19011?

An AS9101D audit has two stages: a *stage 1 audit* and a *stage 2 audit*. It's best to conduct both stages during an internal audit and during an audit of a supplier. Conducting the process approach, customer focus, and prioritizing the audit during stage 1 are keys to the aerospace process approach to audits.

This chapter and Chapter 8 describe the entire flow of stage 1 and stage 2 audit processes. This chapter incorporates material from Chapters 5 and 6 into the stage 1 audit without repeating previously covered content. Some steps of the stage 1 audit are introduced for the first time.

The stage 2 audit flow is described in detail in Chapter 8. Auditor requirements for both stage 1 and stage 2 audits serve as the steps for the auditor to follow.

The stage 1 audit is outlined in the process flowchart shown in Figure 7.1. This review consists of the following steps:

1. Obtain materials for stage 1 audit
2. Process focus
    a. Confirm supporting functions and determine process responsibilities. Study the scope.

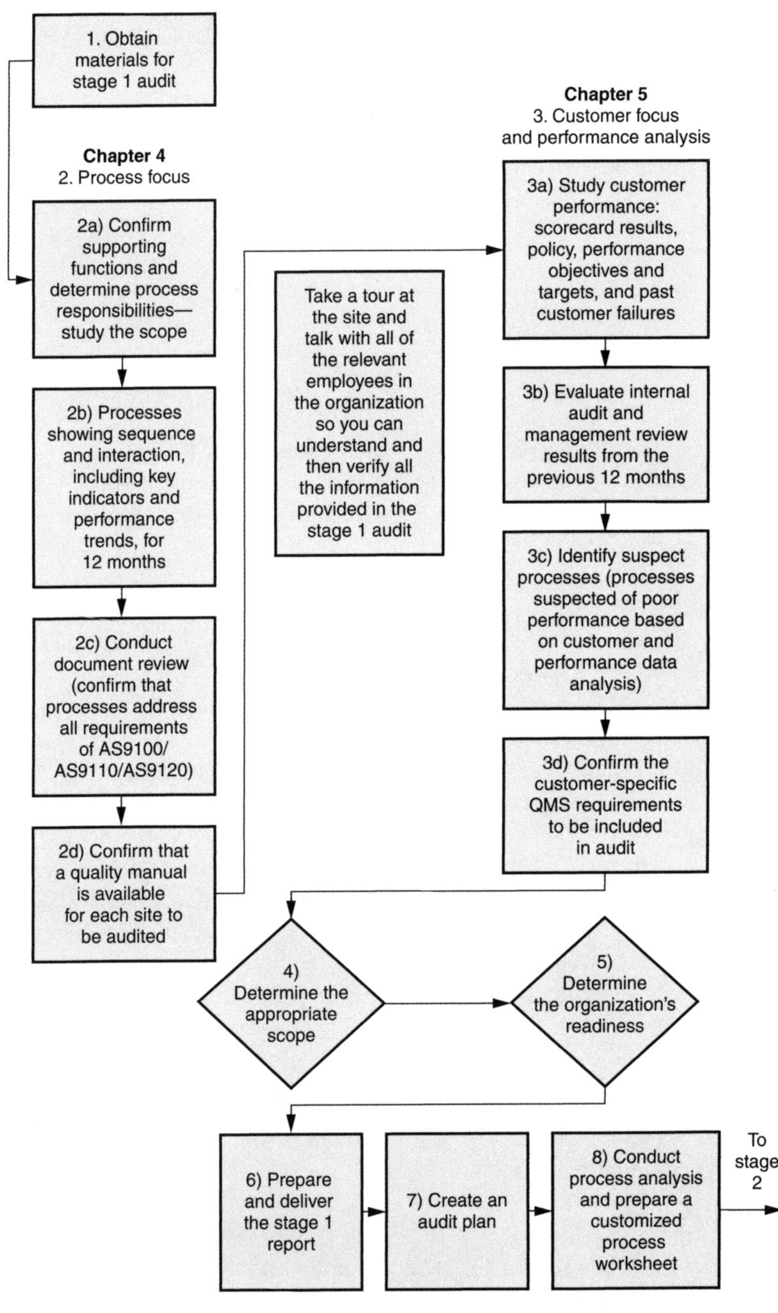

Figure 7.1 Stage 1 audit process flowchart.

   b. Identify processes showing sequence and interaction, including key indicators and performance trends, for 12 months.

   c. Conduct document review and confirm that the processes address all requirements of AS9100/AS9110/AS9120.

   d. Confirm that a quality manual is available.

3. Conduct customer focus and performance analysis.

   a. Study customer performance: scorecard results, policy, performance objectives and targets, and past customer failures.

   b. Evaluate internal audit and management review results from the previous 12 months.

   c. Identify suspect processes (that is, those suspected of poor performance based on customer and performance data analysis).

   d. Confirm the customer-specific quality management system requirements to be included in the audit.

4. Determine the appropriate scope.

5. Determine the organization's readiness.

6. Prepare and deliver the stage 1 report.

7. Create an audit plan.

8. Conduct process analysis and prepare a customized process worksheet.

See Appendix A for the audit report used to conduct a stage 1 audit.

Each step of the stage 1 audit, including the particular steps to be taken by auditors, will be discussed in this chapter.

STEP 1: OBTAIN MATERIALS FOR STAGE 1 AUDIT

During step 1, the auditor secures documentation for the audit (see Figure 7.1). AS9101D clause 4.3.2.2 states that the organization shall provide the following documentation to the audit team for review and use in planning the audit:

- Quality manual

- Description of processes showing the sequence and interactions, including the identification of any outsourced processes
- Performance measures and trends for the previous 12 months
- Evidence that the requirements of the applicable 9100-series standards are addressed by the organization's documented procedures established for the quality management system (for example, by referencing them in the quality manual or by using a cross-reference)
- Interactions with support functions on-site or at remote locations/sites
- Evidence of internal audits of processes/procedures, including internal and external quality management system requirements
- The latest management review results
- List of all major (for example, top five) aviation, space, and/or defense and any other customers requiring 9100-series standard compliance, including an indication of how much business each customer represents and their customer-specific quality management system requirements, if applicable
- Evidence of customer satisfaction and complaint summaries, including verification of customer reports, scorecards, and special status or equivalent

Auditor requirements are as follows:

- Ensure that all required materials have been assembled and reviewed, as listed above.
- Missing information should be identified, and the organization should be apprised of it.

STEP 2: EVALUATE THE PROCESS FOCUS

There are four parts to step 2, evaluating the process focus* (see Figure 7.1):

* See Chapter 5 for a detailed discussion of process focus.

Step 2a: Confirm Supporting Functions and Determine Process Responsibilities—Study the Scope

- *Determine the scope of the audit.* As step 2a in Figure 7.1 indicates, the auditor must begin by understanding the scope of the audit. This involves identifying the site, all its support functions, and all outsourced processes. The scope, products, and processes applicable to an AS9100-series audit should be fully established and studied. At this point, the auditor is asking questions to establish where the support functions are and how they are related. Internal auditors can audit the whole system including aerospace/defense/space and non-aerospace/defense/space products without restricting themselves to the scope requirements of third-party auditors.
- *Ask for organizational charts.* Study all the locations and ask about each site and the different functions associated with it. Identify the site, its design function, purchasing profile (that is, direct and indirect), and top management. Next, identify all the sales offices. Omnex recommends that all sales offices be audited. Identify warehouses and where lab testing takes place. These activities meet the requirements of auditing the site and its support functions.
- *Study the quality manual and process map.* Through interviews, identify all outsourced processes (for example, product and/or process design, plating, and/or heat-treating). Any process that the organization needs for its quality management system and that the organization chooses to be performed by a third party is considered an outsourced process. The auditor should identify each and ensure that these processes are included in the organization's QMS. The auditor should also study the way the organization controls these processes during documentation review.
- *Identify audit responsibilities for the site and its support functions.* Will one audit team be auditing the site, or are multiple auditors responsible? How will the process links and audit trail be managed? If there isn't an audit team responsible for the entire audit process, an agreement should be reached between the different auditors about handing off the auditing processes that connect the site to its support functions.
- If only one audit team is responsible, then it will plan the sampling and auditing of processes that link the site and its support functions whether these are on the same site or at a remote location.

Step 2b: Processes Showing Sequence and Interaction, Including Key Indicators and Performance Trends, for 12 Months

Step 2b in Figure 7.1 focuses on key indicators and performance trends for the previous 12 months. During the audit, the auditor should study the organization's process map or equivalent. Is the map location-specific, and does it explain the processes in the organization being audited? As mentioned earlier, many processes connect between the site and remote locations. In other words, objectives are set by top management. Business planning, objectives deployment, management reviews, new product development, purchasing, and sales are examples of processes that can overlap functional and/or geographical areas. Sample the process documents. Are the interfaces of the processes clearly identified between locations, or do the documents stop within the four walls of the site or support function? (See Figure 7.2.)

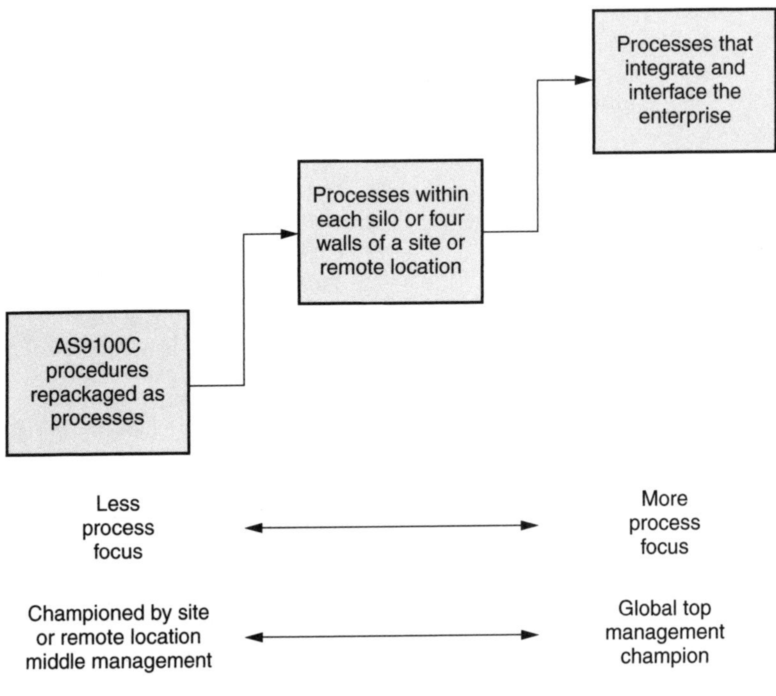

Figure 7.2 Organization of processes by location.

Study the Process Map

There are several other things to look for when studying the process map:

- It should be simple and at the same time descriptive enough to show the sequence and interaction.
- The processes depicted on the map are meta-processes that flow into several lower-level processes typically documented in the level II procedures.
- Does the organization have a process map that shows how all the entities link together and how overall processes link corporate, sales, design, manufacturing, assembly, and the warehouse?
- Does the process map show the sequence and interaction of the processes at the site or entity?
- Study the links or process interfaces for multiple processes between the site and remote locations as well as within the entity being audited. Do the inputs and outputs match? Does the process interface make sense relative to the process being studied? In Chapter 6, three audit trails were introduced that will help the auditor evaluate links and samples that can check process interfaces.
- Be careful about process map paradigms—process versus clauses, and process versus functions. The auditor should be aware of a process approach versus a clause, departmental, or functional approach to developing processes (see Figure 7.2). Processes identified by the organization shouldn't be repeats of the clauses in the AS9100 series. They also shouldn't be departmental or functional processes. Figure 7.3 shows an elemental approach, and Figure 7.4 shows a functional approach.

The process map shown in Figure 7.3 is predominantly clause-oriented. Its monitoring and measuring, resource management, FAI, product realization, communication, internal audits, and data analysis elements simply correspond to clauses of the AS9100 series. Furthermore, the process map shows no interactions and thus fails to document how the organization actually operates.

However, this map does illustrate the "quality paradigm" of professionals who have worked with standards since the days of the MIL standards. Many organizations have used procedures developed by quality professionals based on standards requirements from the 1980s. This quality

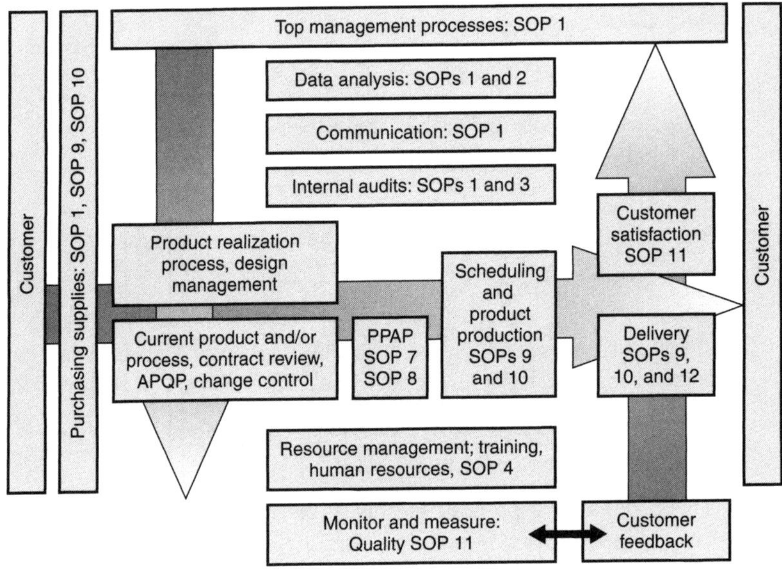

Figure 7.3 Elemental process development approach.

paradigm makes it impossible for implementers or auditors to see beyond the requirements (of standards) to the processes that allow the organization to function.

Process performance must also be examined. Evaluate the process performance data. According to ISO 9001:2005, clause 3.2.14, *effectiveness* is the extent to which planned activities are realized and planned, and results achieved; according to clause 3.2.15, *efficiency* is the relationship between the results achieved and resources used.

There's a big difference between product/process performance and business performance (key indicators and performance trends). The latter is focused more on business results than the former is.

If an organization hasn't maintained 12 months of performance data, it is a major nonconformance in the stage 1 audit.

Figure 7.4 shows a process map that's based on the departments and functions of the plant. Instead of documenting processes, it shows the many departments in the plant, such as engineering, quality, information technology (IT), human resources, and production. Processes aren't functions or departments, but they do cut across them. This map also fails to indicate interactions between its various elements.

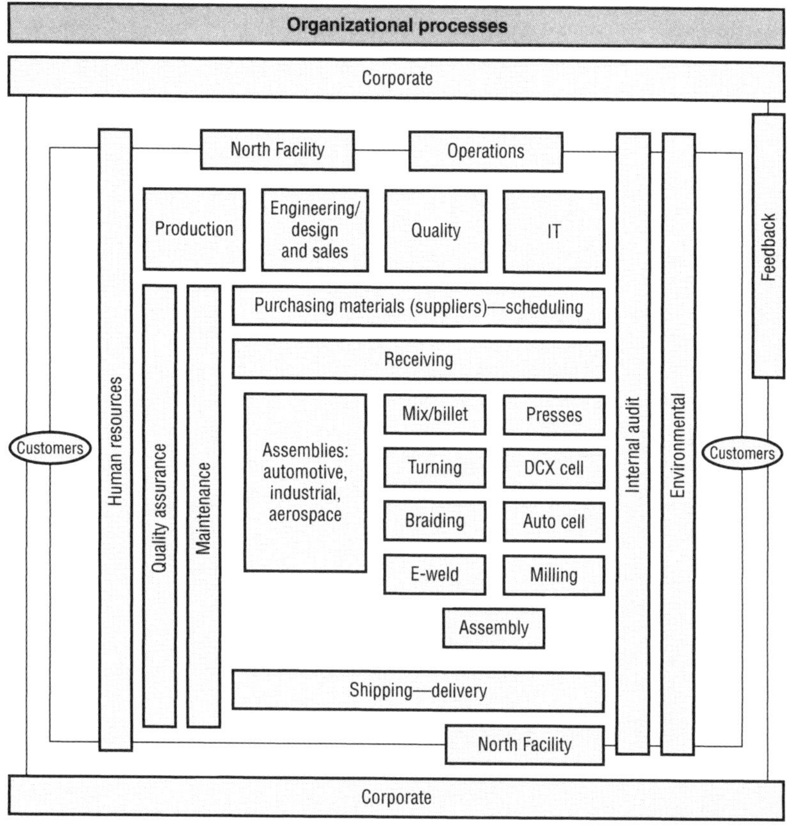

Figure 7.4 Functional process development approach.

Step 2c: Conduct Document Review and Confirm That the Processes Address All Requirements of AS9100/AS9110/AS9120

Study the product realization processes and the COPs and their support processes. Which management processes enable them? Does the process list address all the clauses and subclauses of AS9100? The internal auditors should have the auditee complete a matrix to identify all the process documentation for a particular clause to help the auditor complete the document review easily. To this purpose, this book has edited the QMS process matrix (Appendix D in AS9101D) to include a matrix of subclauses and their associated processes. This edited matrix can be found in the stage 1 and stage 2 checklists (Appendixes A and B). Such a matrix can be completed by the auditee prior to the audit and then verified by the auditor. (*Note:* This is a high-level check only. During the document review, the auditor will look at processes in detail.)

The auditor can move on to the next phase of evaluating the process approach only by studying the processes in detail. Determination of whether a process focus exists can be gained by studying level II process documentation. Keep in mind the following when examining the documentation:

- Study each "shall" or requirement of AS9100/AS9110/AS9120 as well as any customer-specific quality management system requirements. Are they addressed in the process documentation? The documentation should clearly indicate whether the organization is accomplishing or implementing a requirement. (*Note:* Auditors shouldn't accept documents that quote the standard verbatim.)
- If the documentation doesn't explain how the organization addresses a "shall," ask the auditee to show you how it's addressed.
- Level II process documentation should describe what the process does, who does it, and when it's done. The process steps must also be defined.

Note: A process isn't a clause or element of AS9100/AS9110/AS9120, so process descriptions or work instructions that read like one aren't acceptable. A process probably won't be confined within an organizational department, and might not even be confined within the building.

When reviewing a process document, look for the following:

- Process characteristics (for example, process responsibilities and responsible functions are assigned; process sequencing and interactions are established; processes are measured against requirements and defined measures; availability of resources and information; training and effectiveness are monitored, measured, and analyzed against planned results (process effectiveness); actions are implemented to achieve planned results and to promote continual improvement; records are maintained.)
- Are COPs documented with clear inputs from the customer and outputs going back to the customer?
- When studying supplier management, product realization, or other activities that involve outsourced processes, make sure that the organization has exercised control over them.

Step 2d: Confirm That a Quality Manual Is Available for Each Site to Be Audited

Study the quality manual. Evaluate the scope statement, process map, and detailed processes to assess their applicability to the site whose readiness is being evaluated. There's a good chance that the process map is a generic one for the entire organization, and it may not be specific or applicable to the scope of the site and its supporting functions.

At a minimum, the quality manual must contain three things:

1. Scope of the QMS, including details of any exclusions (see step 2a)
2. Documented procedures or references to them (see step 2c)
3. Description of the interaction between the processes of the QMS (see step 2b)

The scope in the quality manual must be identical to the scope of the QMS implemented. The description of the interaction between processes is typically documented using a process map. However, an organization is free to document the processes and their interaction in any way it chooses. (See the description of step 2b earlier in this chapter.)

STEP 3: CONDUCT CUSTOMER FOCUS* AND PERFORMANCE ANALYSIS

Step 3 comprises four parts:

Step 3a: Study Customer Performance: Scorecard Results, Policy, Performance Objectives and Targets, and Past Customer Failures

- When studying customer performance through customer scorecard results, policy, and performance objectives and targets, pay close attention to the following:
  - Does the organization have an overall process for customer feedback (customer feedback and customer satisfaction)? *Note:* This isn't the contract review process. Does this process show

* See Chapter 5 for more on customer focus.

all the interactions with the customer, including actions taken to meet customer requirements and objectives?

- Identify the list of prioritized customer requirements/ expectations. Does it include cost, quality, delivery, and other key customer issues? Were all relevant interfaces with the customer and data used to collect and analyze the list of expectations?
- Has the organization gathered set objectives based on customer expectations? To help determine this, complete the customer expectations sampling sheet shown in Figure 7.5.
- How are the data analyzed from customer feedback made actionable, that is, as action plans and assignments? Do the measures include product conformity and on-time delivery performance to the customer?

- Is there a process in place to gather and analyze customer satisfaction (that is, perception) and customer scorecards?
  - AS9100 requires customer satisfaction evaluation that includes—at a minimum—product conformity, on-time delivery, customer complaints, and corrective action requests. What do the data show regarding customer satisfaction? How well does the organization address customer satisfaction issues?
  - From the list of customers, identify which ones provide scorecards. For each customer, review the customer scorecard online. Carefully identify which customers provide scorecards by plant and which ones score by product line. Don't be satisfied by reviewing printouts because they could be misleading if the customer provides updates electronically.
  - Identify all customer performance issues. Particularly, identify whether the customer has put them into a customer-specific organizational approval status category for improvement. For each customer, ask the organization how it evaluates overall satisfaction. *Note:* Don't confuse scorecard performance with customer satisfaction (perception).
- Determine whether the organization is reviewing and using scorecards internally.
  - What is it doing for problem areas? Are actions taken to address customer issues? Study the action plan and actual

Customer Expectation Sampling Sheet

Customer expectations* (5.2)	Objectives (5.4.1)	Deployed objectives or departments (5.4.1)	Related processes (4.1 or 7.0)	Strategy to meet objectives (5.4.2)	Evidence of objectives being met (5.6 or 8.5.1)
Delivery*	100 percent on time	Purchasing: Supplier shipment 100 percent on time Scheduling: No change in schedule two weeks out ——— Quality: N/A	• Supplier development process • Supplier rating • Scheduling process	• Project to increase uptime from 84 percent to 95 percent • Project to increase supplier short shipments	• No improvement seen. Project initiated two months ago. • Supplier short shipment shows improvement of 25 percent
Reduce product lead time*	Reduce lead time from twenty-four months to sixteen for major product launch	Purchasing: Involve suppliers up front. Scheduling: Reduce supplier submission misses to zero ——— Quality: none	• Project planning • Design strategy process	• Project to reuse technology • Project to increase simultaneous engineering	• Reuse technology in process • Improvement of 20 percent is evident ——— • Simultaneous engineering project implementation in process • No requirement evident

* Are you sampling product quality, delivery, technology, lead-time, reliability, and cost?
* Note: Which dimensions are important to the customer?

Figure 7.5 Customer Expectation Sampling Sheet example.

implementation. Is it meaningful, and is the organization taking actual steps to improve satisfaction?

  - Analyze customer supplemental data, customer scorecards, management review performance data, and customer complaints. What are the key customer performance issues? Identify the suspect processes that affect performance.

- If the supplier is on a special status with its customer, then the supplier could be deemed unready for the readiness review of third-party audits. Bring this to the attention of top management in the closing meeting and the internal audit report.
- Are there any open customer complaints?
  - If yes, has a corrective action plan been implemented or planned?
  - Do any problems repeat, indicating issues with the problem-solving process?
  - Search for patterns or trends among common problems for the same part.
  - Evaluate several problem-solving reports to determine the quality of the problem-solving effort.

Step 3b: Evaluate Internal Audit and Management Review Results from the Previous 12 Months

Evaluate Internal Audit and Management Review Results

When evaluating internal audit and management review results, pay close attention to the following:

- Are internal audit and management review results for the previous year available?
  - An internal audit should include at least one full system audit covering all process map processes and clauses.
  - Ensure that the audit was conducted based on the aerospace process approach to auditing. Review the readiness review and prioritized audit plan prepared by the internal auditors.
- Was the internal audit conducted using a process approach versus an elemental or clause approach?
  - Evaluate the competency of the internal auditors.

  - Evaluate the quality of written nonconformities. Are the findings superficial or performance-based (that is, training and document control versus missed deliveries or customer complaints)?
  - Look for a three-part nonconformity statement.
- Evaluate the closeout process. Does it include both corrective action of the root cause and systemic corrective action to prevent the problem from occurring again?
- Study the yearly audit schedule. Does it show the schedule for system, manufacturing, and/or process audits?
  - Does evidence indicate that the audits are scheduled based on data from past audits, customer concerns, and internal rejects?
  - Is there evidence to show that the schedule is adjusted periodically, based on data?
- Was a management review conducted after the internal audit?
  - Was top management apprised appropriately on the health of the QMS?
  - How often is the management review held? Is it held frequently enough to drive organizational improvement? *Note:* Sometimes, management review is held once a year just to satisfy the AS9100/AS9110/AS9120 requirements. If so, it should be considered insufficient in terms of meeting the requirement.

Clause 5.6.2, Review Input

- Results of audits
- Customer feedback
  - Customer scorecards, and so on, must be included in the management review as customer feedback. *Note:* See the process for customer feedback to ensure that all feedback is reported in the summary to top management in the management review.
- Process performance and product conformity
- Status of preventive and corrective actions
- Follow-up actions from previous management reviews
- Planned changes that could affect the QMS
- Recommendations for improvement

New Product Development Monitoring

- Especially in design organizations, delivery of product can be thought of as "design on-time." Customer satisfaction and customer expectations are "design deliverables on-time." Consider how these are accomplished and how they are reported in the management review of the design group. Also, note that clause 7.3.5, Design Review, requires that there is an authorization for the design to go to the next stage. Who does the authorization and how does it take place? Do problems and risks get reported to top management? How?

Clause 8.2.1, Customer Satisfaction, and Clause 5.6.3, Management Review Outputs

- Are the actions and decisions associated with the management review documented?
- Is top management reacting and taking actions to improve poorly performing management review inputs?
- Is there output from the management review that's action- and improvement-oriented?
- Does the management review cover the minimum requirements of AS9100/AS9110/AS9120?

Step 3c: Identify Suspect Processes (That Is, Those Suspected of Poor Performance Based on Customer and Performance Data Analysis)

Identify Suspect Processes

Next, identify poorly performing processes that create risk for the customer, based on the analysis of customer and performance data. Use the processes identified in this chapter and illustrated in Figures 7.6 and 7.7 to identify the relationship between results and process performance. Prioritize the processes as they relate to the customer and the organizational risk (see Figure 7.7).

Figure 7.6 shows how customer performance problems relate to internal performance problems. Process issues cause external and internal performance issues. These suspect processes need to be identified and included in the prioritized audit plan. Creating a prioritized audit plan is an important step toward conducting an aerospace process-focused audit. (This will be explained in more detail in Chapter 8.)

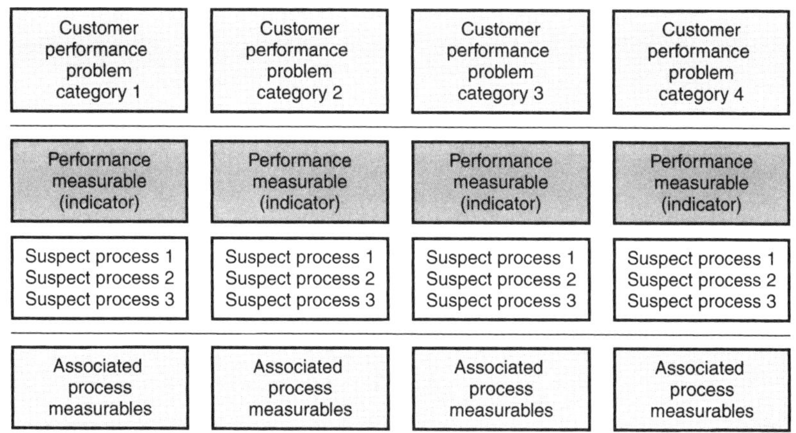

Figure 7.6 Grouping suspect processes.

Step 3d: Confirm the Customer-Specific Quality Management System Requirements to Be Included in the Audit

The auditor needs to identify the complete list of customers. When conducting internal audits where the organization serves multiple markets, the internal auditor should look at all key customers (that is, customers that provide the majority of sales to the organization or site being audited). This list can also include customers the organization intends to target for business growth.

The auditor should verify that the organization is using an effective process to gather and update customer-specific quality management system requirements. The process shouldn't simply identify customer-specific requirements; it should identify changes to the requirements as they occur. By doing this the organization can follow through and make the changes to the appropriate processes.

Examples provided by AS9101D of customer-specific quality management system requirements include product process verification including first article inspection, quality records to be created and maintained, coordination of document changes, defined special requirements, flow-down of requirements, and other requirements; see AS9101D, clause 4.3.2.2.

- The auditor should verify the availability of customer-specific quality management system requirements for all selected customers as applicable.

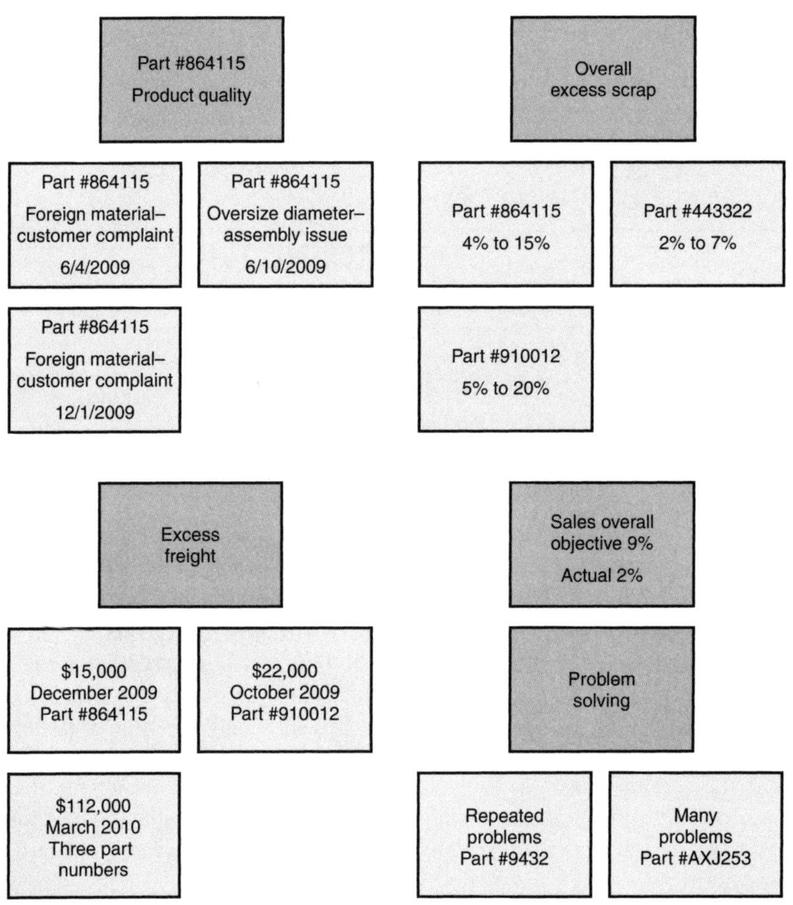

Figure 7.7 Grouping processes by common risk categories.

- Sample a few processes that are affected by particular customer-specific quality management system requirements to ensure that the requirements are implemented at the process level. For example:
  - Flow-down requirements
  - First article requirements
  - Coordination of document changes
  - Customer notification of product and process changes

Two cautions:

- Don't audit customer-specific quality management system requirements as clauses or elements alone.
- Audit as a part of a process.

STEP 4: DETERMINE THE APPROPRIATE SCOPE

When determining the scope during step 4, the requirements in AS9100/AS9110/AS9120, clause 1.2, and AS9101D, clause 4.3.2.4, need to be taken into account.

AS9100/AS9110/AS9120, Clause 1.2

All requirements of this clause are generic and are intended to be applicable to all organizations regardless of type, size, and product provided. Where the requirements aren't applicable due to the nature of an organization and its product, this can be considered for exclusion. Where exclusions are made, claims of conformity aren't acceptable unless they're limited to requirements within clause 7, and such exclusions don't affect the organization's ability or responsibility to provide products that meet customer and applicable regulatory requirements.

Typical clauses that can be excluded include 7.3, Design and Development, especially for those organizations where the customer conducts the design and the supplier performs the manufacturing.

Auditor Requirements

After the auditor has examined the process and customer focus of the organization, the auditor should clearly identify the following:

- Site and supporting functions
- Processes with interfaces between the site and its supporting functions
- Products manufactured

Conduct audits of supporting functions before the audit of the site. This is considered good practice in internal audits as well.

STEP 5: DETERMINE THE ORGANIZATION'S READINESS

In third-party audits, the readiness of the organization to proceed to stage 2 is determined. If the organization isn't ready to conduct the stage 2 site audit, the certification body and the organization can agree to stop the process.

In the internal audit, this is really not a choice. If the organization is not ready, the internal auditors can issue nonconformances that can be added to the final audit report or closed out during the stage 2 audit.

Stage 1 audit results shall be documented and communicated to the auditee, including identification of any areas of concern that could be classified as a nonconformity during the stage 2 audit. The internal auditor should feel free to provide nonconformities in the stage 1 audit, unlike third-party auditors.

The stage 1 audit is the most critical step of the audit, so auditors should be thorough and provide nonconformities as necessary for improvement.

For guidance to internal auditors, the following situations typically may require postponement in a third-party situation or a major nonconformity in an internal audit situation:

- Customer scorecards show that the organization is under a special status category. Some of the customer-specific organizational approval statuses include limited approval, probation, suspension, or withdrawal.
- The organization doesn't have one year of internal audits, management reviews, or performance data.
- An internal system audit to AS9100/AS9110/AS9120 (all processes, clauses, or aerospace process approach to audits) hasn't been completed.
- Management review shows no top management involvement, or the management review is incomplete.
- The organization shows poor customer or process focus.
- Documentation shows many requirements not being addressed by the organization's processes.

If there are obvious major nonconformities with respect to the implementation of the management system, and/or performance or customer issues, the auditor notes them and issues major nonconformities or identifies them in the audit trails for the stage 2 audit. Unlike in a third-party audit, lack of readiness is not a showstopper in the internal audit insofar as the stage 2 is not postponed. In other words, whether the organization is ready or not, the internal auditor will just proceed with the internal audit, issuing nonconformities to the deficient areas.

STEP 6: PREPARE AND DELIVER THE STAGE 1 REPORT

Internal and supplier auditors can identify any major or minor nonconformities and report them in the internal audit.

STEP 7: CREATE AN AUDIT PLAN

The audit plan in AS9100/AS9110/AS9120 should be organized by processes from the process map, not by clause. During step 7, the audit plan should be prioritized according to suspect processes identified in the performance analysis and customer focus step. See the guidance provided earlier in Chapter 5. The auditor should then identify the sequence of processes to audit in the audit plan by referring to the audit trails discussed in Chapter 6: business planning and management (BPM) review, new product development (NPD), and provision audit trails. Finally, the customer oriented processes (COPs) are identified and given priority. *Note:* NPD and provision audit trails are COPs; however, the organization should also have identified additional COPs.

Some good practices to keep in mind:

- Audit based on the defined processes of the organization, not clauses.
- Begin with an audit of top management and cover the following:
  - Scope of the quality management system
  - Processes and their interfaces
  - Criticality of products and processes
  - Product-related safety issues
  - Internal audit results and previous findings
  - Performance measures and trends
  - Management review and actions

- Areas of risk to the customer, including customer complaints and customer-specific organizational approval status notification
- Requirements, including customer-specific quality management system requirements, statutory and regulatory requirements
- Performance data from customers
- Changes to the organization
- Required competencies of audit team members
- Progress toward continual improvement (as related to objectives set)
- Effectiveness of corrective actions, and verification since previous audit

• Include supporting functions, on-site and remote, focusing on interfaces.

• Cover all shifts.

• Include customer-specific quality management system requirements as they apply to sampled processes.

• Review the effectiveness of the QMS in achieving customer and organizational objectives.

• Don't audit as a team except for opening, top management, audit, and closing meetings.

Clause-Driven Audit Plan versus Process-Driven Audit Plan

The clause-driven audit plan is planned around the AS9100 series clauses while a process-driven audit plan shows the organizational processes from the process map. See Figure 7.8.

The real difference between the clause-driven versus process-driven audit plan isn't the audit plan, but the intent of the auditor. The intent of the auditor in the clause-driven audit plan is to ensure compliance to the standard. The intent of the auditor in the process-driven audit plan is to examine the performance of the organization and its processes. In the aerospace audit, the auditor is determining existence of the process approach and conformance to the standard via the OER report.

Each on-site audit (initial, surveillance, and recertification) shall include an audit of:

Clause-driven audit plan		
Date and time:		
Day one:		
8:00 a.m.	Opening meeting	Top managers and management
8:30 a.m.	Plant tour	
9:00 a.m.	Customer focus (5.2)	Marketing manager
11:00 a.m.	Document control (4.2.3)	Quality department
12:00 p.m.	Lunch	

Process-driven audit plan		
Date and time:		
Day one:		
8:00 a.m.	Opening meeting	Top managers and management
8:30 a.m.	Plant tour	
9:00 a.m.	Customer satisfaction and performance data review (8.2.1)	Top management
11:00 a.m.	Operational review (5.6)	Top management
12:00 p.m.	Lunch	

Figure 7.8 Clause-driven and process-driven audit plans.

- Requirements of new customers implemented since last audit
- Customer complaints and organization response
- Organization internal audit and management review results and actions
- Progress made toward continual improvement targets
- Effectiveness of the corrective actions, and verification since the last audit
- Effectiveness of the management system with regard to achieving both customer and organization objectives

Every audit should include auditing on all shifts.

Auditor Requirements

- Study the process map of the organization and identify the three audit trails—BPM, NPD, and provision (see Chapter 6).

- Audit the BPM processes and top management in the beginning of the audit.
  - Start with customer expectations, customer satisfaction, and the scorecards. Review with top management their thoughts and ideas on key issues from a customer viewpoint.
  - Identify which auditor is more proficient in each audit trail, and update the audit plan with the processes related to BPM, NPD, and provision audit trails.
  - Identify all suspect processes and COPs in the organization and make sure they're in the audit plan.
  - Complete the process-to-clause checklist, and make sure that all clauses are being audited.
  - Identify OEM and other customer-specific quality management system requirements. Make notations in the audit plan to ensure that customer-specific requirements will be sampled.
- The auditor should keep in mind the following:
  - Analysis of actual or potential risk to the customer, product, and processes.
  - Auditor/multiple auditors demonstrating links between audit trails.
  - All shifts and manufacturing activities shall be audited where they occur.
  - The required number of audit days.
  - Optimizing audit time based on the layout of the organization.
  - Interfaces with the remote support functions, that is, those audited by other internal auditors.
  - Audit team competencies and language skills, including the use of translators where necessary.
  - Customer-specific quality management system requirements and risks in all relevant processes, including new customers since the last audit.
  - Customer concerns and/or complaints, special status notification, and the organization's response.
  - Internal audit and management review results and actions.

- Progress made toward continual improvement.
- Effectiveness of the corrective actions, and verification since the last audit.
- Effectiveness of the management system with regard to achieving both customer and organizational objectives.
- Issue the audit plan to the organization and all audit team members.

STEP 8: CONDUCT PROCESS ANALYSIS AND PREPARE A CUSTOMIZED PROCESS WORKSHEET

The process analysis in step 8 can be conducted ahead of time (before the audit), or it can be completed when the person is being interviewed (see Figure 7.1).

Preparing the Process Audit Worksheet/PEAR

The auditor should use the Turtle diagram in Figure 7.9 and the Process Audit Worksheet/PEAR in Figure 7.10 for auditing all processes. The Turtle analysis allows the auditor to evaluate inputs, outputs, resources, measurement, monitoring, and methods. The performance analysis and Turtle analysis should assist the auditor in the readiness review in developing process-related questions in the process worksheet. This book provides process checklists in Appendix A to aid the auditor.

Note: Omnex recommends that internal auditors use the Process Audit Worksheet/PEAR to audit all processes of the organization included in the audit plan. The audit plan processes should reference the clauses of the standard (see Figure 7.8).

The internal auditor can use the AS9100C/OER clause-based checklist (Appendix C) to augment the audit of the process to ensure all requirements related to AS9100 are met in addition to process effectiveness audited by the Process Audit Worksheet/PEAR.

Preparing Audit Checklists

When conducting internal audits in the aerospace industry, the auditor can use both an audit checklist like the one presented in Appendix C and the Process Audit Worksheet/PEAR. The checklist is used to evaluate the

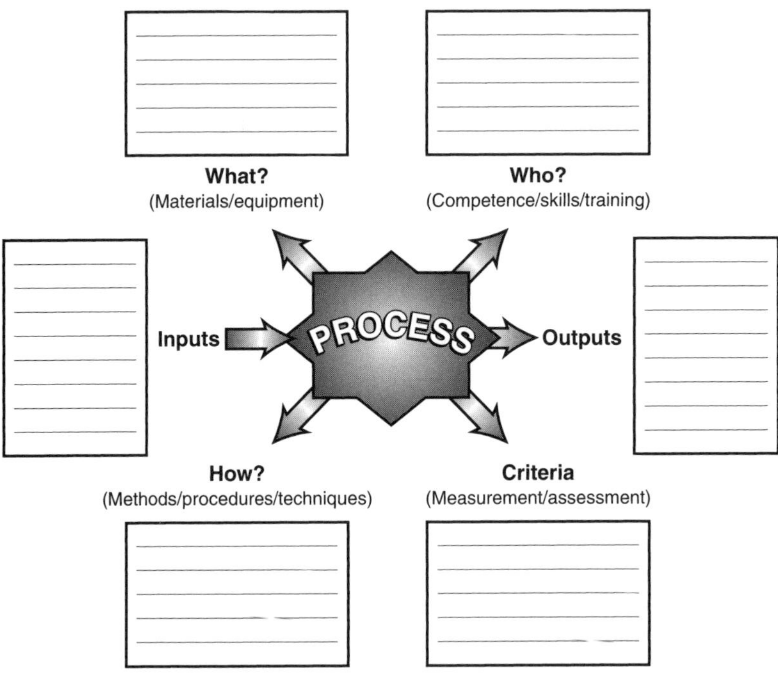

Figure 7.9 Turtle diagram example.

requirements of the standard and document objective evidence, and the worksheet is used to evaluate process performance.

Auditor Requirements

Here's a step-by-step guide for completing the Turtle diagram in the process assessment sheet (O):

- First, identify inputs and outputs. These have to be actual inputs. For example, in management review they're the inputs for clause 5.6.2—or metrics reviewed in the management review—and the output is the action plan. Teams typically add qualifications, such as a well-run meeting, in the output. This is incorrect. The inputs and outputs should be actual hard inputs and outputs.
- After the hard inputs and outputs are identified, the team systematically discusses each leg of the process analysis.

Process Audit Worksheet/PEAR

Company name: Ⓐ	Location: Ⓑ	Audit type: Ⓒ	Standard:
Process characteristics: Ⓓ ❑ Has responsibilities/process owner assigned? ❑ Considers customer specifics ❑ Is documented? ❑ Is monitored?		**Classification:** ❑ Needs further research ❑ Nonconformities ❑ Opportunities for improvements	
Auditor: Ⓔ		**Process/area:** Ⓕ	
Organization's method for determing process effectiveness (for example, key indicators or measurements): Ⓖ		**Applicable clauses (AS9100/9110/9120):** Ⓗ	
Auditees/interviewees: Ⓘ **Audit observations and comments supporting process effectiveness determination:**		**Process interfaces (process linkages) and process details:** Ⓙ **Process activities:**	
Statement of effectiveness:	1. ❑ Not implemented; planned results are not achieved. 2. ❑ Implemented; planned results are not achieved and appropriate actions not taken. 3. ❑ Implemented; planned results are not achieved, but appropriate actions are being taken. 4. ❑ Implemented; planned results are achieved. Ⓚ		

Interactions

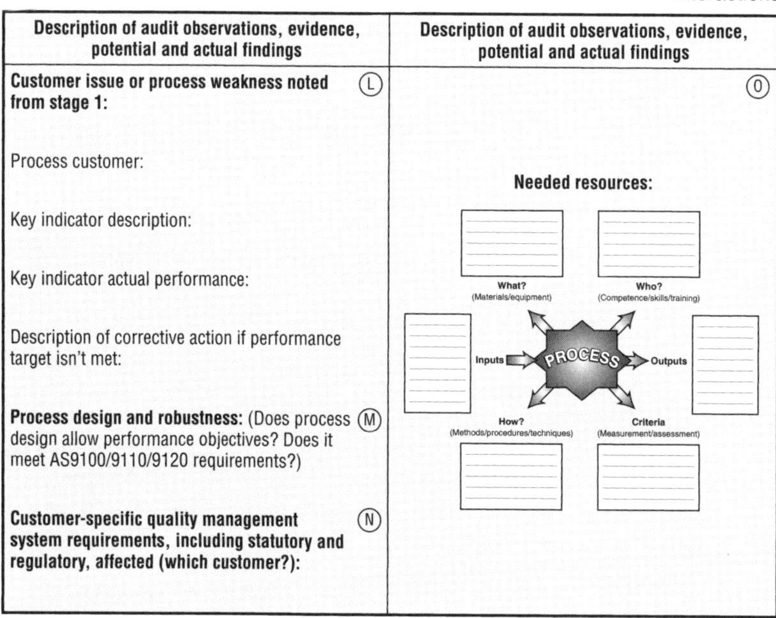

Description of audit observations, evidence, potential and actual findings	Description of audit observations, evidence, potential and actual findings
Customer issue or process weakness noted from stage 1: Ⓛ Process customer: Key indicator description: Key indicator actual performance: Description of corrective action if performance target isn't met: **Process design and robustness:** (Does process design allow performance objectives? Does it meet AS9100/9110/9120 requirements?) Ⓜ **Customer-specific quality management system requirements, including statutory and regulatory, affected (which customer?):** Ⓝ	Ⓞ **Needed resources:**

Figure 7.10 Process Audit Worksheet/PEAR example.

- Analyze skill and competency levels to ensure process success. What kind of competency and skill does management need in order to have a well-run management review?
- The metrics for measuring the process are then discussed. *Note:* These should measure the actual process. They shouldn't be metrics that measure overall results. For example, the process measurement of scheduling shouldn't be on-time delivery, but should be something related to the scheduling process. In management review, the process measurement shouldn't be progress in the measurables, but should be something like the percentage of action items completed.
- The hardware and software required for the process are identified as something like room, projector, or software.
- Finally, the team identifies the procedures, processes, and/or methodology the process uses for overall success.

It's a good idea to complete as much of the Process Audit Worksheet/PEAR as possible for key processes or suspect processes prior to the audit. This includes thinking through the questions that need to be asked. When a suspect or poorly performing process has been identified, it should become the auditor's primary focus.

When completing the process audit worksheet be sure to:

- Complete the header with company name and location (A, B).
- Identify the type of audit—for example, system, process, product, or supplier (C).
- Identify auditor name (E).
- Identify the process and area being audited. Since it is a prioritized audit, processes chosen typically affect customer satisfaction, customer complaints, or organizational performance issues (F).
- Identify which process measurements relating to customer concerns (that is, expectations) should be targeted. Also, add the process measurement that they're presently tracking (G).
- Determine which clauses of AS9100/AS9110/AS9120 the process satisfies. For example, a customer complaint process could have at least two related clauses—8.5.2 and 7.2.3 (H).
- Identify the interfaces or links of the process with other processes. Does the process documentation show the links, and does the

client understand them? You should also summarize the process details (J).

- Determine which customer and customer-specific quality management system requirements affect the process. Also identify all statutory and regulatory requirements (N).
- Prepare the process-related questions to be asked (L):
  - First, identify all the process concerns as they relate to performance issues for the customer and/or overall organizational performance. Does the process explain why there's no overall performance recorded in a customer-related issue, scorecard, or satisfaction survey? If not, identify potential questions relating to supporting processes.
  - Second, identify the general process-related questions, and audit using the Turtle diagram (O). Then determine whether the process is performing satisfactorily and whether it's measured, monitored, and improved as required.
  - Third, identify potential questions that haven't been asked relating to conformity with AS9100/AS9110/AS9120 requirements. This can be documented in the AS9100C checklist and Objective Evidence Record (OER) (see Appendix C).

Not all areas of a Process Audit Worksheet/PEAR can be assessed in the stage 1 audit:

- Process characteristics and classifications (D) are determined by the interview in the stage 2 audit.
- The organization's method for process effectiveness (G) is completed during the interview in stage 2. In other words, how is effectiveness measured and maintained? Typically, the process is measured by process indicators or other measurements. See Chapter 2, Figure 2.9 for the BMS control plan.
- Auditor observations and comments supporting process effectiveness determination (I) refer to objective evidence supporting the auditor's conclusion for the statement of effectiveness (K).
- Statement of effectiveness (K) ranges from "not implemented; planned results are not achieved," to "implemented; planned results are achieved."

- Process weaknesses noted for stage 1 (L), although partially completed during the stage 2 audit, still need additional information from the auditor.
- Process design and robustness (M): if the process repeatedly fails, the auditor needs to ask whether the process has been satisfactorily designed and whether it will need a redesign in order to meet the process objectives.
- The Turtle diagram (O) can be used during the interview to audit the process.

Clause 4.3.2.4, Stage 1 Conclusions

The stage 1 audit completes all the relevant requirements of AS9101D for the internal audit, including:

- Determining the quality management system implementation status.
- Identifying any areas of concern that would be classified as a nonconformity, if not resolved before the stage 2 audit.
- Developing a customer-focused and prioritized plan for the stage 2 audit that includes any additional quality management system requirements from the organization's aviation, space, and defense customers.

The customer and process focus is evaluated in the stage 1 audit. Also, customer and business performance is studied in order to prioritize the audit and the audit plan. The process approach starts during the stage 1 audit.

8
Stage 2: On-Site Audit

CONDUCTING THE AUDIT

All audits, including internal audits, are required to follow the guidelines provided in ISO 19011:2002. Because these requirements are covered in a standard internal or lead auditor course, they're not discussed here. This book covers the AS9101D and related requirements *not* found in ISO 19011:2002 for the stage 2 on-site audit.

The stage 2 flowchart is shown in Figure 8.1. The steps for the stage 2 on-site audit are as follows:

9. Conduct audit of remote supporting functions (recommended)
10. Opening meeting
11. Conducting the audit:
    a. Conduct facility tour, if needed
    b. Study customer and organizational performance
    c. Meeting with top management
    d. Audit organizational processes
    e. Verify that all processes and clauses are audited
12. Write up nonconformities
13. Closing meeting:
    a. Determine audit team next steps
    b. Prepare the draft report
    c. Conduct the closing meeting

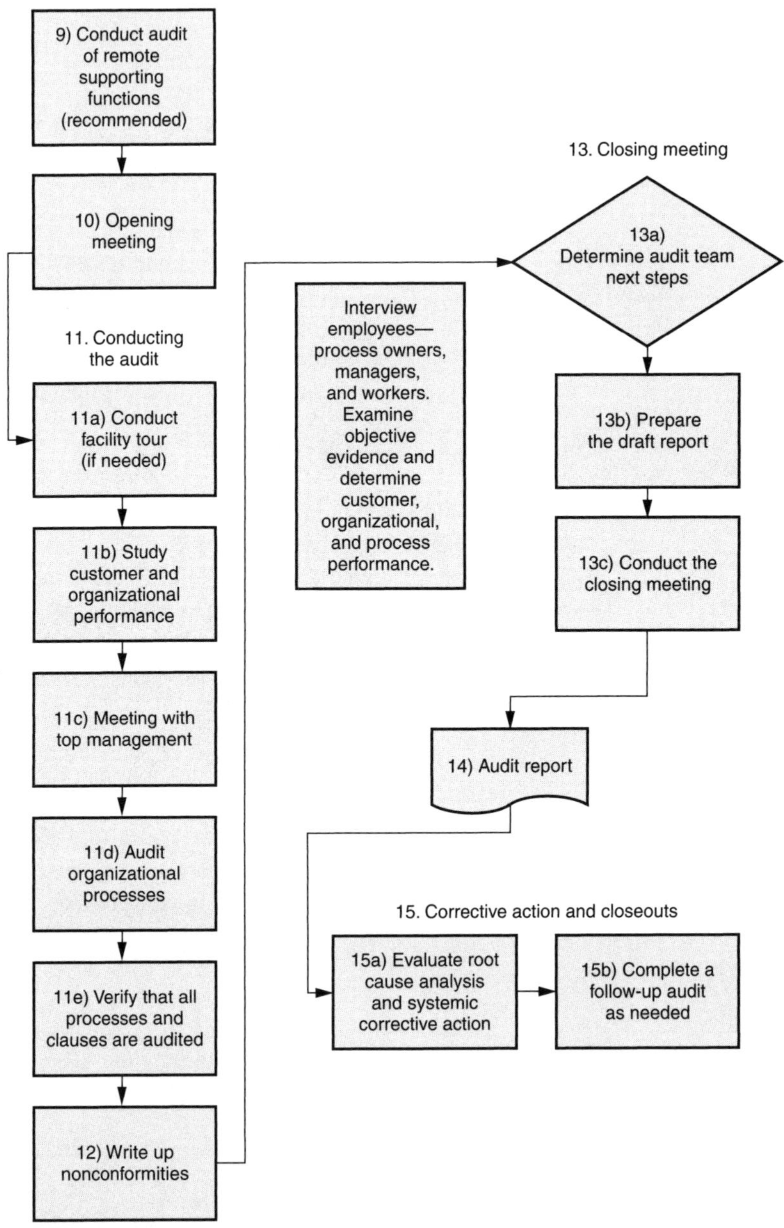

Figure 8.1 Stage 2 audit process flowchart.

14. Audit report
15. Corrective action and closeouts:
    a. Evaluate root cause analysis and systemic corrective action
    b. Complete a follow-up audit as needed

This book has covered in some detail the customer-focus and process-focus requirements of an audit (see Chapter 5). After the stage 1 audit the auditor should continue following the prioritized audit plan, taking into consideration the customer and performance issues noted during the on-site audit. The auditor also needs to determine if the organization actually reflects the process-oriented nature of its process map and process documentation of the stage 1 readiness review.

STEP 9: CONDUCT AUDIT OF REMOTE SUPPORTING FUNCTIONS

Step 9 in Figure 8.1 is just a reminder that the supporting functions should be audited first as best practice before the on-site audit is conducted. All the steps of the stage 2 audit should be completed for these functions before the audit of the manufacturing site.

Many of the processes are initiated at the supporting functions and remote locations. The samples taken at the supporting functions can be followed to the site whether they're sales, design, purchasing, or top management functions. Auditors need to keep in mind that they're starting at the supporting functions in order to check the interfaces between processes—whether these involve deploying objectives or developing new product.

Auditor Requirements

- Identify the processes that provide a link between support functions and the site. These processes should be evident in the process map itself. Are they?
- Does the process documentation connect the site to the remote location?
- How is the process managed? Is it working effectively?
  - Take samples for each of the processes that will be used to test the interface. Whether they involve top management objectives, new products in sales and/or design, or integrating customer

satisfaction information into the sales office, carefully sample them and follow each process from the supporting function into the site. Does the process link to other processes? How is it measured, monitored, and improved? How is the process managed?

- Audit the process using stage 2 step 11d.
  - Are the supporting functions and sites working together? Are they designed to succeed together?

STEP 10: OPENING MEETING

During step 10, the opening meeting, the auditors are introduced to the organization's management team (see Figure 8.1). This gives auditors an opportunity to explain the overall aerospace auditing process. ISO 19011:2002 has detailed requirements for the opening meeting.

Auditor Requirements

During the opening meeting, the auditor needs to inform the client of the following additional items important to aerospace auditing:

- Aerospace process approach.
- Audit plan.
- Reconfirm the following:
  - Customer satisfaction and complaint status, including customer reports and scorecards
  - Supporting functions and interfacing processes
- Revise the audit plan based on any new information provided since the stage 1 audit.
- Confirm the following:
  - The audit will encompass a cross-section of the organization: top management, management, workers, and engineers.
  - The audit will take place where the work is done and not in an office. Even if the records are computerized, the audit should be done in the cubicle of the engineer or the manager, as required.

– Large crowds can't follow the auditor. These tend to skew the audit process. Why? Auditees need to be in a natural setting and made to feel at ease for the audit to be most successful.

– The auditor will immediately point out nonconformities to the auditee.

See the opening meeting checklist in Figure 8.2, which encompasses both the ISO 19011 and AS9100/AS9110/AS9120 requirements.

Opening Meeting Checklist

- Introduce audit team and attendees.
- Pass out the attendance sheet.
- Explain the aerospace process approach.
- Review objectives, scope, and criteria.
  - Review supporting functions and interfacing processes.
- Summary of methods and procedures used for audit:
  - Auditor takes notes of details for both conformities and nonconformities.
  - Audit conclusion is based on samples taken.
  - The audit will encompass a cross-section of the organization, from top management to maintenance workers and engineers.
  - The audit is restricted to small groups of three or fewer people.
  - The auditor notifies auditee of nonconformities during the audit as well as during a daily review meeting.
  - Questions should be directed toward lead auditor.
  - Conditions are specified for when a major nonconformity affecting the customer is uncovered.
- Establish auditee communication link.
- Reconfirm the following:
  - Current customers
  - Customer satisfaction and complaint status, including customer reports and scorecards
  - Any customer special status (bad supplier status)
  - Supporting functions and interfacing processes
- Review prioritized audit plan.
- Confirm status of stage 1 issues, including documentation.
- Confirm time and date of closing meeting.
- Confirm relevant safety, emergency, and security procedures.

Figure 8.2 Opening Meeting Checklist.

STEP 11: CONDUCTING THE AUDIT

When conducting the audit in step 11, the auditor needs to follow all the points made during the opening meeting (see Figure 8.2). In other words, audit a cross-section of the organization, especially top management. Audit time should include and involve top management, especially for the business planning and management review audit trail.

Involving the process owners in each of the interviews regarding the process during stage 2 is important. The process owner should be able to explain the beginning and the end of the process, and how it is monitored, measured, improved, and resourced.

An example of poor process ownership can be seen in an audit performed in a large Fortune 100 organization. An engineering change was being audited in this organization. The manager explained to the auditor that the engineering change was done using an ERP software called SAP, and she introduced herself as the process owner. She was asked how many of the change requests originated with the customer. She replied with mild irritation that she took care of the changes and didn't know about how the process started and how the data were input into the system. After a while, the auditor asked her whether changes on the production lines that ensued were noted or not. She replied with a clear flash of anger, "How would I know what happens in manufacturing with these changes?"

Clearly, in this case, the process owner couldn't explain the start and end of the process. And even worse, the auditor discovered that this person couldn't explain the start and end of the process because there was *no one* in charge of the engineering change process.

The auditor should follow the prioritized audit plan by targeting poorly performing processes as they relate to overall performance. When auditing a process, the auditor is trying to discern what's wrong with it and whether process performance is the cause of the poor customer performance or poor overall performance. Next, the auditor performs process analysis using a Turtle diagram and process characteristics to discern the process definition.

Audit Trails and Audit Planning

The prioritized audit plan has already considered three main audit trails, including business planning and management (BPM) review, new product development (NPD), and provision. The key to these audit trails is the choice of samples that the auditor follows. On the BPM audit trail, the auditor follows the links of customer expectations from objectives, to processes, to continual improvement. (See Figure 6.3 for an example of the Customer Expectations Sampling Sheet.) The NPD audit trail samples "contracts" from sales, to new product development, to the first article inspection (FAI) or product and process validation. The provision audit trail follows the process flow of a part family using an inspection sheet and key characteristic data.

When the audit is being conducted, the auditor should follow side trails to investigate a problem. Remember, the auditor is investigating customer issues, customer satisfaction, or performance issues. When the auditor is in the design department and an issue of training comes up, the auditor should follow that issue to the human resources department either at that moment or later when the training process is audited. As another example, if while examining the nonconformity crib the auditor finds many samples noted as "design engineering samples" that have little to no identification, he or she is free to immediately follow this trail to the design department.

Note that the audit shouldn't take place in a conference room, but at the location of the auditee. Also, the auditor should audit the person or item being sampled while keeping in mind two issues: relevancy and representative sampling.

First, the sample taken has to be relevant within the scope of the AS9100-series audit being conducted. This is an important topic for discussion in a third-party audit; for example, in an aerospace, defense, or space audit, an agricultural product being supplied by the auditee is not a good sample and is not relevant to the scope of the audit. In an internal audit, the scope is not as strict, and the auditor is free to pursue samples beneficial to the organization, unlike the third-party auditor. Relevancy in an internal audit may be more important in keeping discipline to the samples taken for a particular audit topic. For example, when auditing management review, the internal auditor should stick to samples important to management review without suddenly jumping to design review or some other subject matter. In other words, design review is not a relevant sample for the topic of management review.

The idea of representative sampling is important when auditing any area of the organization (for example, sampling purchase orders in purchasing). The auditor can randomly sample between steel suppliers A, B, C, and D and choose four purchase orders from each. However, in the audit this would be a nonconformity against the auditor. Why? The auditor didn't first understand the process. If the auditor had asked the right question, he or she would have found out that 80 percent of the steel is purchased from supplier D, 10 percent from A, and five percent each from C and D. Understanding this, the auditor should have selected samples based on the process.

The second application of representative sampling is in manufacturing. The auditor should ideally sample the key manufacturing processes as well as the product, as per the "vital few" parts in the organization.

This same logic should be followed while sampling personnel within the organization. Top management should be audited for the BPM audit trail and processes. For all others, the process owners and those working in the processes should be sampled. The auditor should keep in mind the personnel working in the organization and what the statistical breakdown is for top management, managers, engineers, quality staff, and workers on the plant floor.

Following the Prioritized Audit Plan

When conducting the aerospace process approach to audits, the priority of the AS9100/AS9110/AS9120 audit is to uncover risks to the customer and the organization. The auditor should sample personnel and records to also check conformance; auditing performance isn't enough. The auditor should complete each process audit, keeping in mind that enough samples should be covered to verify that the process is implemented and that it conforms to the organization's documented process and the relevant AS9100-series standards.

Conducting an Aerospace Process Approach to Audits

Process monitoring and improvement (audit method introduced in Chapter 6 on audit trails) is the method used to audit all processes. The forms and method provided here will hold the auditor in good stead when auditing processes using the aerospace process approach.

The QMS contains three types of processes (see Figure 8.3):

- Customer oriented processes (COPs)
- Management oriented processes (MOPs)
- Support oriented processes (SOPs)

Links and Samples

As explained in AS9101D, the auditor is required to prioritize the audit based on the organization's performance. Study examples of customer feedback, including nonconformity data, corrective action requests, results of satisfaction surveys, and complaints regarding product quality, on-time delivery, service provision, or responsiveness to customer and internal requests, including OASIS. Try to relate these failures to the processes that are not performing satisfactorily. The process map comes in handy when

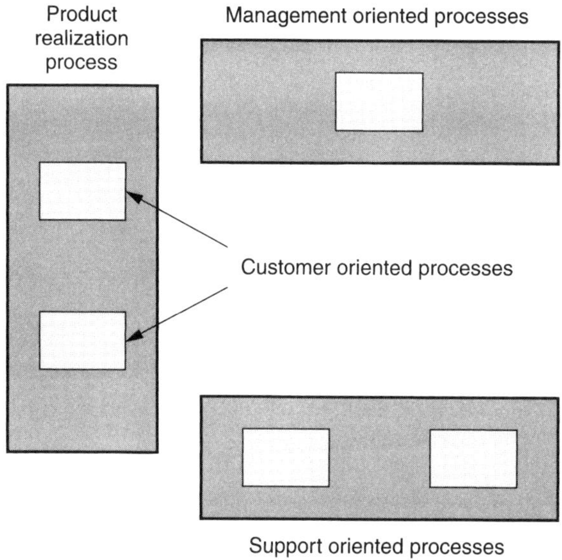

Figure 8.3 Quality management system process map.

identifying which processes to sample. This phase in the aerospace process approach to audits is conducted as step 3b in the readiness review (see Chapter 5).

Next, the auditor should conduct a process analysis using a Turtle diagram to identify sources of potential variation in the process. This is step 3c in the stage 1 audit (see chapter 7). The key to an effective process analysis is to determine the actual inputs, outputs, and customer expectations. In the example in Figure 8.4, for the management review process the inputs to the business reviews are the agenda items that will be covered during the meeting. The output is the meeting minutes and action plan. Expectations for the outputs are fully attended meetings and actions. The expectations constitute the criteria to determine effectiveness of the process.

Once this is done, the auditor should determine the methods and/or procedures, material and/or equipment, and the competence, skills, and training required for the process to be effective. The process analysis (Turtle diagram) can be conducted prior to or in conjunction with the audit. It will show any weakness in the process definition.

Although the auditor has initiated the analysis during the readiness review, he or she needs to interview the auditee (that is, process workers) to determine what's actually happening in the process.

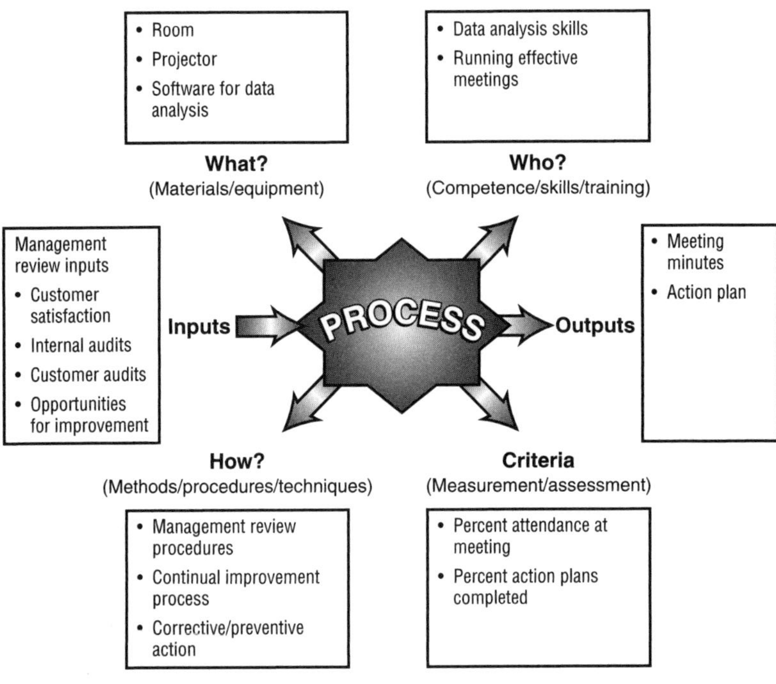

Figure 8.4 Turtle diagram example.

In the example in Figure 8.4, if the Turtle diagram for management review showed a key problem identified from the performance analysis (for example, lack of actions), then this would be an area the auditor would explore further when conducting the audit. Questions based on the performance review and process analysis should be documented on the Process Audit Worksheet/PEAR prior to the audit, as discussed in step 8 of the stage 1 audit. During the audit, these questions should become the focus of the effectiveness review or Process Audit Worksheet/PEAR of the audit. In other words, risks to the customer and performance should be the focus of the Process Audit Worksheet/PEAR.

Completing the Process Audit Worksheet/PEAR

A Process Audit Worksheet/PEAR such as the one shown in Figure 8.5 should be used to record the objective evidence identified during the performance effectiveness auditing. One of the key purposes of the worksheet is ensuring that process characteristics of a process have been addressed

Process Audit Worksheet/PEAR

Company name: Ⓐ	Location: Ⓑ	Audit type: Ⓒ	Standard:
Process characteristics: Ⓓ ❏ Has responsibilities/process owner assigned? ❏ Considers customer specifics ❏ Is documented? ❏ Is monitored?		**Classification:** ❏ Needs further research ❏ Nonconformities ❏ Opportunities for improvements	
Auditor: Ⓔ		**Process/area:** Ⓕ	
Organization's method for determing process effectiveness (for example, key indicators or measurements): Ⓖ		**Applicable clauses (AS9100/9110/9120):** Ⓗ	
Auditees/interviewees: Ⓘ **Audit observations and comments supporting process effectiveness determination:**		**Process interfaces (process linkages) and process details:** Ⓙ **Process activities:**	
Statement of effectiveness:	1. ❏ Not implemented; planned results are not achieved. Ⓚ 2. ❏ Implemented; planned results are not achieved and appropriate actions not taken. 3. ❏ Implemented; planned results are not achieved, but appropriate actions are being taken. 4. ❏ Implemented; planned results are achieved.		

Interactions

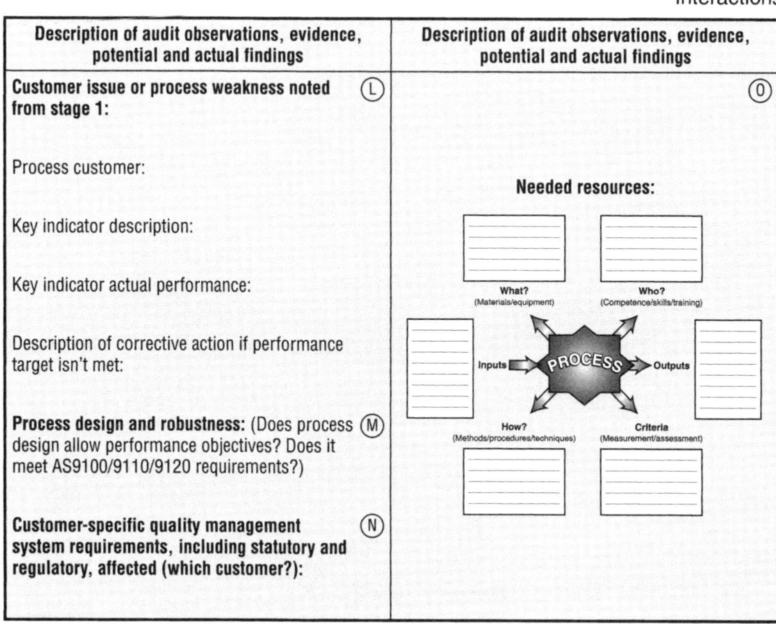

Description of audit observations, evidence, potential and actual findings	Description of audit observations, evidence, potential and actual findings
Customer issue or process weakness noted from stage 1: Ⓛ Process customer: Key indicator description: Key indicator actual performance: Description of corrective action if performance target isn't met: **Process design and robustness:** (Does process design allow performance objectives? Does it meet AS9100/9110/9120 requirements?) Ⓜ **Customer-specific quality management system requirements, including statutory and regulatory, affected (which customer?):** Ⓝ	Ⓞ **Needed resources:**

Figure 8.5 Process Audit Worksheet/PEAR example.

(see Figure 8.6). The process characteristics are placed on the header of the Process Audit Worksheet/PEAR and the Turtle diagram at the very end.

The auditor can use the process characteristics to audit processes that are chosen from the process map (see Figure 8.7). when conducting the stage 2 audit.

It's a good idea to complete as much of the Process Audit Worksheet/PEAR as possible for key, suspect, and customer oriented processes (COPs) prior to the audit. This includes thinking through the questions that need to be asked. When a suspect or poorly performing process has been identified, the customer performance issue should become the auditor's primary focus.

When completing the process audit worksheet be sure to:

- Complete the header with company name and location (A, B).
- Identify the type of audit—for example, system, process, product, or supplier (C).
- Identify auditor name (E).
- Identify the process and area being audited. Since it is a prioritized audit, processes chosen typically affect customer satisfaction, customer complaints, or organizational performance issues (F).
- Identify which process measurements relating to customer concerns (that is, expectations) should be targeted. Also, add the process measurement that they're presently tracking (G).
- Determine which clauses of AS9100/AS9110/AS9120 the process satisfies. For example, a customer complaint process could have at least two related clauses—8.5.2 and 7.2.3 (H).

A process can be identified by a series of unique but consistent characteristics. Eight characteristics of a process are recommended for effective (quality) management:

1. Responsibilities assigned (process name and owner)
2. Process activities
3. Process limits
4. Process control methods
5. Process operation methods
6. Needed resources
7. Applicable AS9100 requirements
8. Applicable customer, regulatory, and statutory QMS requirements

Figure 8.6 Process characteristics.

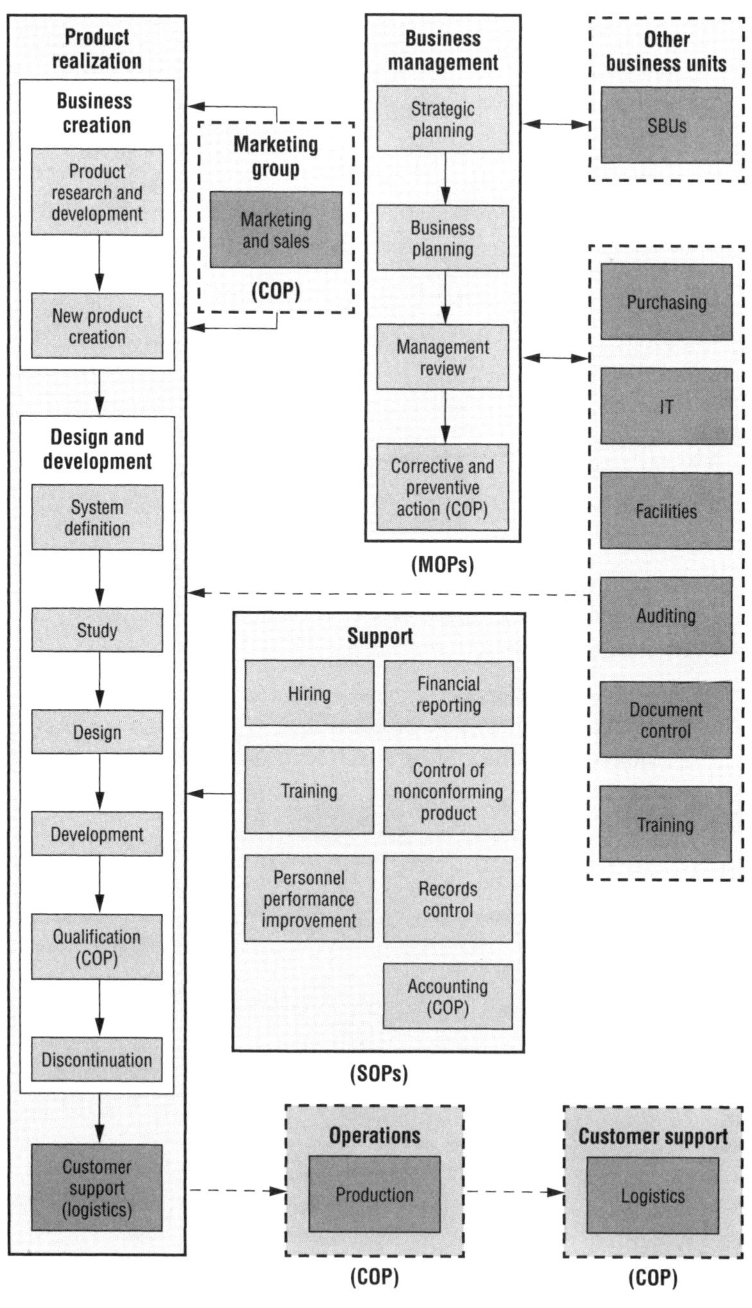

Figure 8.7 Process map example.

- Identify the interfaces or links of the process with other processes. Does the process documentation show the links, and does the client understand them? You should also summarize the process details (J).
- Determine which customer and customer-specific quality management system requirements affect the process. Also identify all statutory and regulatory requirements (N).
- Prepare the process-related questions to be asked (L):
  - First, identify all the process concerns as they relate to performance issues for the customer and/or overall organizational performance. Does the process explain why there's no overall performance recorded in a customer-related issue, scorecard, or satisfaction survey? If not, identify potential questions relating to supporting processes.
  - Second, identify the general process-related questions and audit using the Turtle diagram (O). Then determine whether the process is performing satisfactorily and whether it's measured, monitored, and improved as required.
  - Third, identify potential questions that haven't been asked relating to conformity with AS9100/AS9110/AS9120 requirements. This can be documented in the conformance/OER checklist (see Appendix C).

Not all areas of a Process Audit Worksheet/PEAR can be assessed in the stage 1 audit:

- Process characteristics and classifications (D) are determined by the interview in the stage 2 audit.
- The organization's method for process effectiveness (G) is completed during the interview in stage 2. In other words, how is effectiveness measured and maintained? Typically, the process is measured by process indicators or other measurements. See Chapter 2, Figure 2.9 for the BMS control plan.
- Auditor observations and comments supporting process effectiveness determination (I) refer to objective evidence supporting the auditor's conclusion for the statement of effectiveness (K).
- Statement of effectiveness (K) ranges from "not implemented; planned results are not achieved," to "implemented; planned results are achieved."

- Process weaknesses noted for stage 1 (L), although partially completed during the stage 2 audit, still need additional information from the auditor.
- Process design and robustness (M): if the process repeatedly fails, the auditor needs to ask whether the process has been satisfactorily designed and whether it will need a redesign in order to meet the process objectives.
- The Turtle diagram (O) can be used during the interview to audit the process.

Performance Analysis

The aerospace audit is meant to add value. It should start with a performance analysis of customer data that identifies areas of weakness or areas for improvement as noted in the readiness review. It should end by identifying variations (that is, nonconformities) in the process that, if eliminated, would lead to process improvement.

Identify a process weakness and probe that weakness during the audit. Start by asking questions about performance as it relates to the overall weakness in customer scorecards, rejections, or other key performance failure observed:

- Identify what's expected. What are the indicators, objectives, and actual performance?
- How is performance being improved?
- How was the process planned (M)? Does the process design allow performance objectives to be met? Does the process design meet the requirements of the AS9100 series being audited? What about robustness? Does the process take into account previous performance results?
- Follow the process using the organization's documented process flow or procedure.
- Is the process being carried out as designed? Are the methods being applied? Sample the process, as applicable, where the work is done (that is, engineering, shop floor, or workstation).

After the process approach audit is complete, the auditor should identify whether the process is sufficiently defined, working satisfactorily, relates to overall performance, and includes well-established links. The auditor should sample the process to verify that it's working according to the process definition.

Also, this is a good time to rate the overall effectiveness of the process:

1. Not implemented; planned results are not achieved.
2. Implemented; planned results are not achieved and appropriate actions not taken.
3. Implemented; planned results are not achieved, but appropriate actions are being taken.
4. Implemented; planned results are achieved.

The evaluation is fairly straightforward from "not implemented" to "implemented." If it is implemented, effectiveness comes into play for ratings 2, 3, and 4. If the planned results are not met, the organization gets a 3 if there is a suitable action plan, or a 2 if it does not. The best possible rating for a process is a 4, and it indicates that the process is implemented and achieving planned results.

Auditors should realize that completing the Process Audit Worksheet/PEAR and OER are key steps in conducting the audit. The auditor must ensure that the samples taken are documented. Objective evidence of the audit should be noted, as well as conformance and nonconformity issues. This is an important point for many auditors who might not document all the issues found during the audit. When conducting an audit, the auditor needs objective evidence that shows whether a process is working well or poorly.

Step 11a: Conduct Facility Tour, If needed

Step 11a, the facility tour, can be conducted as necessary by an internal auditor visiting from another location, or second-party auditor when visiting a site or remote location for the first time (see Figure 8.1). Sometimes, a facility tour can be quite revealing. For example, I completed the opening meeting in a plant that claimed it wasn't responsible for design. During the plant tour, I was casually shown a small design group for a part going to a second-tier customer. For whatever reason, the plant had forgotten to mention this function at this site.

The plant tour can be used to confirm the information provided by the organization and the audit plan organized by the lead auditor. Sometimes, issues affecting the audit surface during the tour. It's a good idea to hold a caucus meeting after the tour to discuss what each auditor saw and how it's going to be handled.

Auditor Requirements

Auditor requirements for step 11a include the following:

- Modify the audit plan based on information collected during the opening meeting and facility tour.
- Identify the need for translators, if necessary.
- Use the project worksheet to write down issues or areas to investigate.

Step 11b: Study Customer and Organizational Performance

There may be up to 30 days or more between the readiness review and the on-site audit. During the interim, the auditor needs to reexamine the customer satisfaction scorecards and organizational performance, per step 11b (see Figure 8.1).

Auditor Requirements

Auditor requirements for step 11b include the following:

- Reexamine the customer scorecard. Has performance maintained the same level, or has it deteriorated?
- Study the last business performance review. Has the same level of performance been sustained? Did the organization follow up on the actions identified in the management review provided during the readiness review?
- Has the organization been put on a customer-specific organizational approval status notification?
- Consider whether to adjust or reprioritize the audit plan based on the latest customer and performance issues.

Step 11c: Meeting with Top Management

Step 11c, the meeting with top management, includes a number of obligations from AS9101D. These are, by clause: 5.1, Management Commitment; 5.2, Customer Focus; 5.3, Quality Policy; 5.4, Planning; 5.6, Management Review; 8.2.1, Customer Satisfaction; and 8.5.1, Continual Improvement:

- *Clause 5.1, Management Commitment.* This requirement obligates top management to direct, enable, and accomplish certain tasks.
- *Clause 5.2, Customer Focus.* This requires top management to understand customer expectations and help meet them.

- *Clause 5.3, Quality Policy.* Top management must help ensure that a quality policy is established. In clause 5.4.1, Quality Objectives, management must also set objectives that align with that quality policy.
- *Clause 5.4, Planning.* Top management must set measurable objectives that are deployed within the organization. It also must create a plan to meet those objectives.
- *Clause 5.6, Management Review.* Top management participates in reviewing data required by ISO/TS 16949:2002 to help the organization improve and change.
- *Clause 8.2.1, Customer Satisfaction.* Requires top management to review customer satisfaction and to make decisions and take actions to improve it.
- *Clause 8.5.1, Continual Improvement.* Management review is a required component in continually improving the effectiveness of the quality management system.

Management Responsibilities That Can Be Delegated

Certain top management activities can be delegated, and some can't. When the word "ensure" is used in AS9100, as it is in clauses 5.2, Customer Focus, 5.3, Quality Policy, and 5.4, Planning, those items can be delegated.

Auditor Requirements

As discussed earlier, top management's responsibilities are key to a successful quality management system (QMS) implementation. Moreover, AS9101D has specific requirements for top management to fulfill. Interview top management to learn the following:

- Establishment and continued relevance of the organization's quality policy and objectives
- Establishment of performance measures aligned to quality objectives
- Quality management system development, implementation, and continual improvement
- Top management commitment
- Quality management system performance and effectiveness
- Performance to customer expectations (for example, supplier rating, scorecard, audit results)

- Actions taken to address issues/areas that are not meeting customer performance expectations

After the interview, how well did the top manager answer the relevant questions asked regarding the QMS and its performance, and also about customers and the organization's success in meeting customer satisfaction?

The auditor is asked to document his or her observations and record them.

This audit of top management should take place within the context of the processes for the BPM audit trail.

Step 11d: Audit Organizational Processes

Auditor Requirements

Auditor requirements for step 11d are as follows:

- Demonstrate the use of the process approach, customer focus, and leadership:
  - Prioritize audit plan based on risks to customer (see Chapter 7).
  - Use of an OER and Process Audit Worksheet/PEAR as primary tools for auditing processes.
  - Customize the checklist with issues that affect the customer, affect meeting customer-specific quality management system requirements, and/or appear to be poor process definition issues.
- Use the audit plan and the organization's defined processes, including sequence and interaction. Don't be requirements oriented; be process oriented.
- Be able to determine whether the organization is operating according to its process definition, that is, a process map and its sequence and interactions (see Chapter 4). Do the processes reflect reality?
- Audit processes to determine if each one is capable of meeting the key process indicators and is performing satisfactorily.
- Make sure that the customer-specific quality management system requirements are identified, addressed, and maintained in the business management system. (*Note:* Customer-specific requirements should be integrated into the processes.)
- Conduct interviews about the process with those who are involved in the process at its location. Avoid conference room audits. Always

take samples or have the auditees show objective evidence when interviewing them to confirm their statements. The auditor should always choose the samples to avoid biased samples being given.

- Document both conformity and nonconformity in the Process Audit Worksheet/PEAR. The information should be clear enough for an independent review by a third party, if necessary.

Step 11e: Verify That All Processes and Clauses Are Audited

The auditor should follow the audit plan and ensure that the audit encompassed all processes and clauses. Complete the stage 2 process/clause matrix.

STEP 12: WRITE UP NONCONFORMITIES

Nonconformities should be written during the audit as it progresses. Also, any nonconformities should be communicated to the auditee each day as findings. (See step 12 in Figure 8.1.)

Nonconformities are a starting point for continual improvement and as such are an important part of the audit. Auditors feel apologetic about nonconformities. They feel bad about finding issues with the organization's processes. This can cause auditors to avoid fundamental issues. For example, an organization that's conducting a management review only once a year is probably doing so only to satisfy AS9100/AS9110/AS9120 rather than using it as part of its actual improvement process. A second company might not be following its own new product development process. A third company could be doing a poor job solving problems. The auditor should give each of these organizations a major nonconformity. Auditors should ask themselves whether they avoid giving nonconformities due to a lack of knowledge, or reluctance to confront challenging organizational issues.

Another key structural issue noted by auditors is customer focus. Many organizations might not be customer focused for any of the following reasons:

- There is no process for capturing all interfaces with the customer, and consequently no data collection, analysis, and presentation to top management of actions designed to make the organization more customer oriented.
- Setting objectives isn't clearly linked to customer needs and expectations.

- There are no clear actions or analysis of customer perception and scorecards designed to effectively address customer satisfaction or dissatisfaction.

Overall, auditors need to understand that organizations perform in a competitive and complex world. The three audit trails and the customer oriented processes really need to be effective. If they're not, the auditor simply must write nonconformities.

AS9101D has defined major and minor nonconformities as follows.

Major Nonconformity (Clause 3.2)

> A non-fulfillment of a requirement which is likely to result in the failure of the quality management system or reduce its ability to assure controlled processes or compliant products/services; it can be one or more of the following situations:
>
> - a nonconformity where the effect is judged to be detrimental to the integrity of the product or service;
> - the absence of or total breakdown of a system to meet a 9100-series standard requirement, an organization procedure, or customer quality management system requirement;
> - any nonconformity that would result in the probable shipment of nonconforming product; and/or
> - a condition that could result in the failure or reduce the usability of a product or service and its intended purpose.

Minor Nonconformity (Clause 3.3)

> A non-fulfillment of a requirement which is not likely to result in the failure of the quality management system or reduce its ability to assure controlled processes or compliant products/services; it can be either one of the following situations:
>
> - a single system failure or lapse in conformance with a 9100-series standard or customer quality management system requirement; or
> - a single system failure or lapse in conformance with a procedure associated to the organization's quality management system.

All nonconformities need to be recorded but not closed during the audit. The internal auditor and the organization need to require the auditee to submit root cause analysis and evidence of systemic corrective action for each nonconformity issued.

A written nonconformity must include three items (see Figure 8.8):

- Statement of nonconformity (system level)
- The requirement that's not being fulfilled
- Objective evidence

The statement of nonconformity is expressed as an issue with the system and shouldn't be expressed against a person or an incident. The nonconformity statement should define *the system problem.*

The unmet requirement should be quoted. It can be from AS9100/AS9110/AS9120, the customer-specific quality management system requirements, and/or an item within the process documentation of the organization. Auditors should carefully study the requirements of the AS9100-series standard being audited, or the organization's own processes that are being violated, and quote the section not being addressed. In no case should the auditor cite several clauses or many procedure numbers in a single nonconformity. Such a lack of precision from the auditor creates confusion for both the auditor and auditee.

The objective evidence and statement of nonconformity can be written in one or two sentences, maximum. Often, writing a paragraph doesn't make the nonconformity any more clear; it actually makes it more vague.

Auditors should be bold and have confidence in their abilities. They need to evaluate whether the process is working for the organization. Is the company satisfying its customers and market?

Auditor Requirements

- Nonconformity statements must contain a statement of nonconformity against the system, the unmet requirement, and the objective evidence.
- One nonconformity can be written to cover more than one "shall."
- Nonconformities must be categorized as major or minor (see definition above).
- Nonconformities should be cross-referenced to an organization's QMS and/or relevant clause of AS9100/AS9110/AS9120.

Auditor

Statement of nonconformity (express as a system problem):

The problem identification system is not effective.

7.5.3 "The organization shall identify the product by suitable means throughout . . ." —Auditor note

Objective evidence:

Three tubs of parts weren't identified with a product tag in department 350.—Auditor note

Auditee:

Extent of problem:

Spot audits were done by manufacturing personnel on all shifts during the week of 11/15/2006, and they noted problems in many other departments, including 350, 340, 220, and 150.—Auditee note

Containment:

Immediately asked all departments, including 350, to tag all the untagged baskets. 11/16/2006.—Auditee note

System root cause: (use five whys, if applicable):

Operators aren't following tagging procedure. Some supervisors aren't stepping up and informing workers about the need to follow tagging procedures.—Auditee note

- *Why? New employees and supervisors—Auditee note*
- *Why? Tagging isn't included in the new employee orientation training.—Auditee note*

System corrective action and impact (auditee notes):

1. *Train all plant supervisors on the tagging procedure, and ensure that they train operators who work in their departments. 11/30/2006—J. Black*
2. *Conduct two manufacturing process audits a month. The process audits will survey many departments for tagging issues. Change implemented 11/30/2006—M. Hank*
3. *Update new employee orientation training to include tagging by 11/15/2006—D. Barnes*

Auditor Closeout

Verification of corrective action (auditee notes):

- Reviewed training record 10-B for supervisory training held on 12/12/2006. OK.
- Reviewed process audits for 11/2006 and 12/2006. Is implemented and working.
- New employee orientation training updated. Verified 1/15/2007.

Resample problem:

Reviewed tagging in departments in 110, 315, 350 on 1/15/2007. OK—Auditor note

Figure 8.8 Sample evidence of corrective action closeout.

- Identify opportunities for recommendations without offering solutions.
- Use a format that has root cause, corrective action, and systemic action.

STEP 13: CLOSING MEETING

The auditor is responsible for three things during step 13, the closing meeting:

- Determining audit team recommendations
- Preparing the draft report
- Conducting closing meeting

Closeout Meetings

Auditors should be careful not to accept any nonconformity closeouts during the audit. Why? The organization will likely just fix the symptom instead of identifying the root cause and *then* fixing the problem. Can the organization actually implement a significant preventive action, or will this problem repeat again soon? It will probably repeat.

During the closeout, let the auditee know the deadlines for when the corrective action responses are needed.

Step 13a: Determine Audit Team Next Steps

Once all of the nonconformities are written, step 13a focuses on the audit team's recommendations. There are four outcomes:

- Minor and major nonconformities are so numerous that another system audit will be recommended to management.
- Minor nonconformities that can be closed via written documentation.
- Minor nonconformities that require on-site closeout.
- Major nonconformities that require on-site closeout.

Step 13b: Prepare the Draft Report

The minimum content for a draft report is all the nonconformities and the audit team summary. This draft report should be presented during the closing meeting and left with the auditee. (See step 13b in Figure 8.1.)

Auditor Requirements

- Prepare draft reports describing all nonconformities. Also, identify and include the audit team summary, at a minimum.

- Identify nonconformities and opportunities for improvement. No other categories are allowed.

Step 13c: Conduct the Closing Meeting

Auditor Requirements

During step 13c, the closing meeting, the auditor has three responsibilities:

- Conduct the closing meeting using the closing meeting agenda defined by ISO 19011:2002 or the organization itself.
- Deliver the draft audit report.
- Record any problems.

Depending on the quality of the data provided, auditors shouldn't shy away from asking for objective evidence and physical closeouts. Internal audit systems should also define their corrective action submission timelines. (There will be more discussion of closeout requirements later in this section.)

It's also a good idea to have an appeals process for internal and supplier audits similar to that of third-party registrar requirements. The appeals process should begin with the lead auditor and move up to a designated senior management person at the organization.

A typical closing meeting agenda includes the following:

- Statement of thanks
- Attendance list
- Scope, objectives, and criteria
- Significance of audit sample
- Audit standard, rules, and reference manuals
- Audit summary
- Root cause and systemic corrective action responses
- Opportunities for improvement
- Clarification of nonconformity statements and summary
- Statement of confidentiality
- Appeals process
- Follow-up
- Close

STEP 14: AUDIT REPORT

Step 14, the audit report, includes requirements from AS9101D.

The audit report should include:

- Details of the auditee and the auditors:
  - Names of both parties
  - Date and address at which the audit was conducted
- Audit scope and products.
- List of customers and customer-specific quality management system requirements with revision level.
- Summary of the audit:
  - Good systems and/or processes found during the audit
  - Overall conclusion of the audit (for example, 14 nonconformities with two major nonconformities)
  - Systems and/or processes not working well that need to be improved
  - Positive feedback about the organization that may encourage them as they move forward
- Nonconformity summary. This is a detailed summary of the nonconformities identified by topic, area where found, and so on. It's easier to see patterns when this list is analyzed.
- Objective evidence collected. The auditor lists all the evidence sampled for all the processes in this area.
- Attendance lists for the opening and closing meetings. The list of attendees of the opening and closing meetings should be included. It's important that top management participates so that the audit will be perceived as truly valuable to the organization.
- Audit plans. These list all the details of the processes targeted and the persons interviewed.
- Clause-to-process cross-references.
- Nonconformities. The audit report will provide all the nonconformities, identified sheet by sheet.

Formality of the Audit Report

The same type of formal report as a third-party audit should be used for the internal audit. Formality in the audit and the audit report will help the internal audit be treated in the same manner as an external audit. Formality will also give the nonconformities and the closeout a sense of appropriate seriousness.

Clause-to-Process Matrix

The auditor should verify that the organization's processes cover all the requirements of the AS9100 series being audited. Clauses or subclauses not audited signify an incomplete system audit.

Management Representative Acceptance of Audit Report

It's good practice to have a sign-off from the organization's management representative.

Auditor Requirements

- Use the audit report model outlined in this chapter and as shown in Appendix B.
- Get sign-off from the auditee that he or she received the report.

STEP 15: CORRECTIVE ACTION AND CLOSEOUTS

Step 15 is the final and most important step of the audit (see Figure 8.1). It's also the only one where the organization being audited saves money. How does this happen? The savings occur when the auditor identifies customer and performance issues in the organization. The auditor subsequently traces the problem to suspect processes, which are then investigated during the stage 2 audit.

During the investigation, nonconformities are identified for process-related problems or lack of process performance that leads to customer and overall performance issues. For instance, during a recent audit an overall yield issue was traced to process issues in the organization. The plant manager told the auditors that this was the best audit in which he'd participated. He said that the issues identified were important to him and his overall success.

Depending on the issues, when an organization fixes the nonconformities identified by the audit, customer satisfaction will increase and performance problems will be solved. Identifying problems during the closing meeting and in the audit report isn't enough, though. The organization's overall response to audit corrective actions must also improve. Thus, it's helpful for the organization if the auditor identifies the process involved and the audit expectations.

The corrective action and closeout phase consists of two steps with different responsible parties. Corrective action is the responsibility of the auditee, and closeout is the responsibility of the auditor.

Auditee's Responsibility

For the corrective action step, the management representative usually assigns the corrective action to the process owner or the manager of a particular department. The first step the manager needs to take is to investigate the problem further and identify its true extent.

Let's say, for example, that two gages are found by an auditor to be out of calibration. The manager or process owner needs to investigate how many other gages are out of calibration. ISO 19011:2002 and AS9100/AS9110/AS9120 require the manager or process owner to find the root cause and conduct the corrective action. Some schemes go further and ask the organization to identify the system corrective action or preventive action. In other words, could the gage calibration problem occur in other departments, in other products, and/or in other plants? How did this problem occur in the first place, and how can the organization stop the problem from happening again? This, then, is one of the best practices recommended in this book for internal audits in the corrective action and closeout steps.

Auditor's Responsibility

The auditor should conduct follow-up verification to close out the nonconformity. He or she should first ask what the problem is, what the root cause related to the problem is, and whether the corrective action is related to the problem. Did the organization investigate the extent of the problem? Did it determine the systemic root cause? If this first step is OK, the auditor should check to see that the corrective actions taken are implemented effectively. In Figure 8.8, the auditor checks to see if both of the corrective actions are implemented as stated. The auditor needs to record objective evidence that the corrective actions, both containment and system actions, have been carried out. Lastly, the auditor needs to resample the problem to ensure that the problem went away.

Step 15a: Evaluate Root Cause Analysis and Systemic Corrective Action

During step 15a, the audit team turns its attention to root causes. There's a cause-and-effect relationship between a root cause and a problem. The root cause can "turn on" the problem when it's present, and it can "turn off" the problem when it's removed. The team could use brainstorming or "is/is not" analysis to identify the root causes. In most cases, there will be many root causes, but they all fall under one of the following three categories: an occur root cause, an escape root cause, and a system root cause. The *occur root cause* made the problem happen. The *escape root cause* allowed the system to let the problem escape. The *system root cause* allowed the problem to occur in the first place (see Figure 8.9).

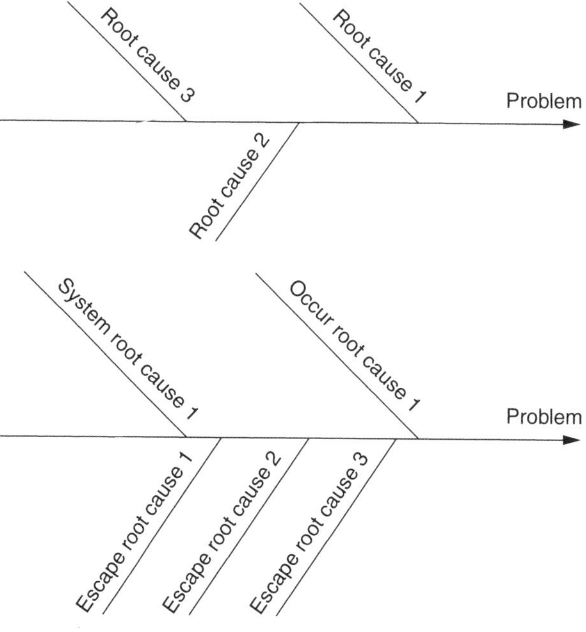

Figure 8.9 Root causes of a problem.

System Corrective Action versus Incident-Specific Corrective Action

If, for example, there was a nonconformity of two tubs not being tagged in department 350, the corrective action could be specific to that department, and the two tubs tagged. This would be called *incident-specific corrective action.*

System corrective action would look across the plant at the product identification process, or look at fixing the problem permanently in the system. For example, instead of just tagging two tubs in department 350, the product identification process should be fixed.

Corrective Action Formats Used in Industry

In 1995 there was a corrective action report (CAR) process widely used across industry. The format was fairly simple, with a problem column and two columns of short- and long-term corrective actions. This was followed by a column for the auditor closeout.

In 1995 Omnex introduced the Eight Disciplines (8D) Problem Solving format for corrective action response for audit closeouts. This is the most popular format used in industry. The "five whys" approach first introduced by Dr. Kaoru Ishikawa refers to the process of going from the symptom to the true root cause. Each time "why" is asked and answered, the true root cause becomes more apparent. Omnex uses this technique in its 8D class.

Evaluating Corrective Action Responses

The following five questions can be used by auditors to evaluate corrective action responses. The questions are applied to the "untagged tubs in department 350" problem referred to earlier. The auditor's responses are shown in italics.

1. *Problem statement:* Has the organization expressed the problem as a system issue? (*Note:* The specific incident identified in the problem isn't the system problem.)

   Yes, it has identified the system problem.

2. *Containment action:* Has the organization taken containment actions on the specific incidents cited by the audit team in its objective evidence? If external customers could be affected, then containment actions must be taken.

   It isn't known if containment action was taken or if the problem is indeed a risk to the customer. In other words, the tagging issue could end up as mixed parts. The answer is therefore not satisfactory.

3. *System root cause:* Has the organization found the root cause(s) of the problem in its system that allowed the incidents cited by the audit team to occur? Has the organization answered the question, "What in our system failed that allowed this to happen?"

   Yes, the system root causes and five whys have been answered.

4. *System corrective action and/or corrective action effect:* Does the corrective action determined by the organization address changes to the system, not specific employees or machines? Does it address the root causes it has identified?

   Yes, it has identified corrective actions to the root causes identified.

5. *Verification of corrective action:* Has the audit team verified that the corrective action is implemented?

   - Major nonconformities need to have an on-site visit for verification.
   - Minor nonconformities may require on-site verification.

   If not, include the verification in the audit plan for the next audit.

   - The decision needs to be made by the auditor, that is, on-site verification or verify at next audit.
   - If the root cause can't be found, or if the corrective action can't be verified, then the nonconformity should be 100 percent resolved.

   Resampling of problem

   - The problem was resampled.

   Overall: the issues identified regarding containment will have to be completed before the nonconformity can be closed out.

Auditor Requirements

- The auditor should study the nonconformity statement and the root causes associated with it. Do the root cause and the problem go hand in hand? Has the organization identified why the problem occurred and why the problem wasn't caught by the QMS? Blaming an operator or someone in the organization isn't an acceptable root cause.

- Study the corrective action associated with each root cause. Will it fix the problem in the system? Study the systemic corrective action. Can the system corrective action prevent the problem from recurring in the future?
- For each corrective action and system corrective action, sample incidents to verify that the action took place. Write the objective evidence in the closeout report.
- Sample the problem again to make sure it has been fixed. Why? Many times, root causes are just guesses of what caused a problem. The only real way the auditor knows the problem has been fixed is if it stops happening.

Step 15b: Complete a Follow-up Audit As Needed

Step 15b is concerned with the follow-up audit (see Figure 8.1).

Omnex recommends a follow-up audit only if there were major nonconformities and/or a number of minor nonconformities for internal audits.

Auditor Requirements

Sometimes, the closeout requires a follow-up audit because of a major nonconformity, risk to the customer, or simply because the closeout must be verified on-site.

There are several options available for closing out a nonconformity:

- The nonconformity can be closed out based on records. (This should be the choice of last resort for internal auditors due to their proximity to the audit location. This is only for minor nonconformities, *not* major nonconformities.)
- The nonconformity has paper evidence and a well-documented plan (see the definition above of 100 percent resolved). The nonconformity can be closed out and rechecked in a subsequent follow-up audit. If there are a number of open issues based on records, the auditor should conduct an on-site audit within 90 days.
- The nonconformity poses a major risk to customers, so it requires on-site verification. However, it will take time to be put into

place. In such a case, the auditor can do an on-site review of the containment actions and then revisit the scenario again during surveillance, or sooner.

CONCLUSION

This chapter has covered the additional requirements of AS9101D and the AS9100-series auditing process. The audit needs to be conducted in a specific way for it to be truly a process approach audit. This chapter, along with the discussion of the stage 1 audit in Chapter 7 and discussion of process worksheets in Chapter 6, defines the process for conducting the aerospace internal audit.

Conducting a monthly audit of a few processes and then calling it a system audit doesn't constitute a system audit. A system audit is a snapshot in time of how an overall system is performing. This includes all processes and all clauses. Conducting monthly audits of clauses or one process at a time won't give the auditor a practical estimation of how the system is working as a whole.

Auditors are encouraged to develop their skills step by step. Becoming an effective auditor doesn't happen by taking one class or reading one book. It requires patience to develop the required knowledge and skills. Figure 8.10 shows the progression.

Risk management tools include:

- Project management and disciplined new product development process
- Risk assessment, characteristics designation, and flow-down
- Design tools, including SFMEA/DFMEA and associated tools
- Process tools, including process flowchart, PFMEA, control plans, and work instructions
- Planning and conducting the FAI (production trial run)
- Understanding and reviewing the FAI
- Team facilitation
- Measurement systems analysis (gage R&R, linearity, bias)
- Statistical process control (C_p, C_{pk})

For more information about the stage 2 audit, see the Confidential Assessment Report in Appendix B.

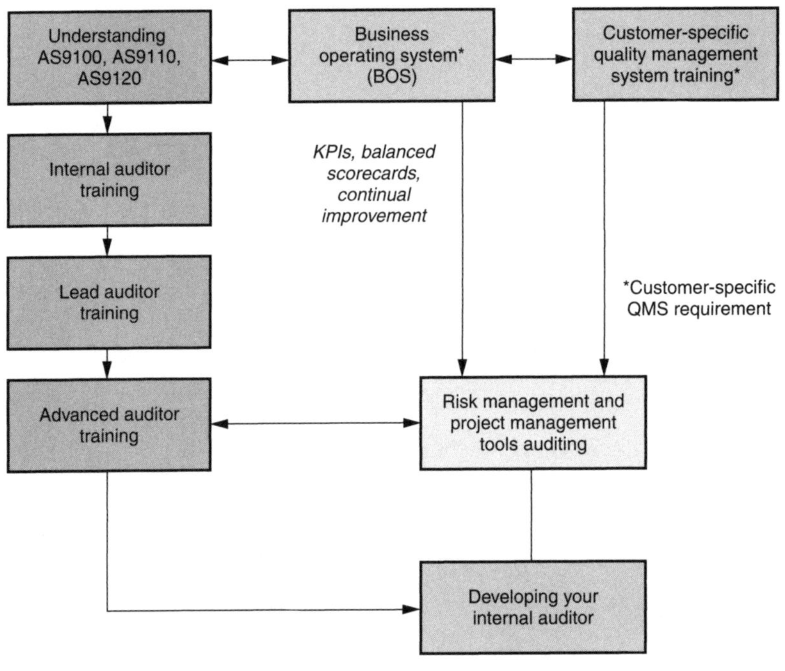

Figure 8.10 What makes a great internal auditor?

Appendix A
Confidential Assessment Report for Stage 1

Organization name: ________________________________

City, state: ________________________________

Organization contact: ________________________________

Lead auditor: ________________________________

CONTENTS

(See Chapter 7 for more details)

Organization Information

Assessment Plan—Stage 1

Step 1: Obtain materials for stage 1 audit.

Step 2: Evaluate the process focus.

Step 2a: Confirm supporting functions and determine process responsibilities. Study the scope.

Step 2b: Processes showing sequence and interaction, including key indicators and performance trends, for 12 months.

Step 2c: Conduct document review and confirm that the processes address all requirements of AS9100/AS9110/AS9120.

Step 2d: Confirm that a quality manual is available for each site to be audited.

Step 3: Conduct customer focus and performance analysis.

Step 3a: Study customer performance: scorecard results, policy, performance objectives and targets, and past customer failures.

Step 3b: Evaluate internal audit and management review results from the previous 12 months.

Step 3c: Identify suspect processes (that is, those suspected of poor performance based on customer and performance data analysis).

Step 3d: Confirm the customer-specific quality management system requirements to be included in the audit.

Step 4: Determine the appropriate scope.

Step 5: Determine the organization's readiness

Step 6: Prepare and deliver the stage 1 report

Step 7: Create an audit plan

Step 8: Conduct process analysis and prepare a customized process worksheet

ORGANIZATION INFORMATION

Organization name: ______________________________

Contact person: ______________________________

Department: ______________________________

Telephone: ______________________________

Fax: ______________________________

E-mail: ______________________________

Street: ______________________________

City: ______________________________

State or province: ______________________________

Zip or postal code: ______________________________

Standard: AS9100C

Type of assessment: Stage 1

Assessment team: ______________________________

Person days: ______________________________

Start date: ______________________________

End date: ______________________________

Assessment information: ______________________________

____

Assessment Plan—Stage 1				
Objective: AS9100C stage 1 audit to ascertain readiness for stage 2 audit				
Date	**Time**	**Auditor**	**Location**	**Process description**
				Opening meeting
				Facility tour
				Review organization structure for support/processes location
				Process focus and document review—quality manual, level I and level II documents
				Customer focus Identify customer-specific QMS requirements
				Customer performance metric review (scorecards) and review of associated corrective actions as appropriate
				Customer complaint status review
				Management review, customer needs and expectations, and organization's performance metric review and review of associated corrective actions as appropriate
				Review internal audit results
				Prepare stage 2 assessment plan
				Closing meeting
				Adjourn

Figure A.1 Assessment plan for stage 1 audit.

Opening Meeting (Process Focus)/Closing Meeting (Prepare and Deliver the Stage 1 Report) Checklist The client will maintain the original meeting attendance list and the original for each audit locally	
Opening meeting discussion	**Closing meeting discussion**
Introduce audit team and members ❑	Review scope ❑
Pass out attendance sheet ❑	Audit standard, rules, and reference manuals ❑
Explain the aerospace process approach ❑	Audit methodology ❑
Review objectives, scope, and criteria ❑	Audit recommendations ❑
Summary of methods and procedures used ❑	Root cause and systemic corrective action response ❑
Establish auditee communication link ❑	Confidentiality ❑
Reconfirm current customers, customer satisfaction and complaint status, and any customer special status ❑	Appeals process ❑
Reconfirm support functions and interfacing processes ❑	
Review prioritized audit plan ❑	
Confirm status of stage 1 issues, including documentation ❑	
Confirm time and date of closing meeting ❑	
Confirm relevant safety, emergency, and security procedures ❑	

Figure A.2 Opening and closing meetings checklist.

STEP 1: OBTAIN MATERIALS FOR STAGE 1 AUDIT

- Study step 1 of the Assessment Plan at the beginning of this report (Figure A.1).
- Ensure that all of the required materials have been assembled and reviewed as required.
- Missing information and documents must be identified, and the organization must be apprised of them.

Required Materials

- ❑ Quality manual
- ❑ Description of processes showing the sequence and interactions, including the identification of any outsourced processes
- ❑ Performance measures and trends for the previous 12 months
- ❑ Evidence that the requirements of the applicable AS9100-series standards are addressed by the organization's documented procedures established for the quality management system, for example, by referencing them in the quality manual or by using a cross-reference
- ❑ Interactions with support functions on-site or at remote locations/sites
- ❑ Evidence of internal audits of processes/procedures, including internal and external quality management system requirements
- ❑ The latest management review results
- ❑ List of all major (for example, top five) aviation, space, and/or defense and any other customers requiring AS9100-series standard compliance, including an indication of how much business each customer represents and their customer-specific quality management system requirements, if applicable
- ❑ Evidence of customer satisfaction and complaint summaries, including verification of customer reports, scorecards, and special status or equivalent

STEP 2: EVALUATE THE PROCESS FOCUS

Step 2a: Confirm Supporting Functions and Determine Process Responsibilities. Study the Scope

The auditor needs to start out by understanding the scope of the audit by identifying the site, all its support functions, and all outsourced processes.

The scope, the products, and the processes applicable to the AS9100 standard audited need to be fully studied and established.

Ask for Corporate Organizational Charts

Study all the organization's locations and ask about each site and the different functions associated with each. Identify the site,* design function for the site, purchasing (direct and indirect), and top management for the site. Next, identify all the sales offices. All the sales offices must be audited as a good practice and/or randomly sampled, at a minimum. Identify warehouses and where lab testing takes place in the organization. This meets the requirements of auditing the site and its support functions.

Study the quality manual and process map. Through interviews, identify all outsourced processes (for example, product or process design, plating, and/or heat-treating). An outsourced process can be any process performed for the organization by a supplier that affects product conformity. The auditor should identify all outsourced processes and ensure that all are included in the organization's QMS. The auditor should make a special note of outsourced processes and study the way that the organization controls these processes during documentation review.

Identify Audit Responsibilities for the Site and Its Support Functions

You must also include the support functions when you identify responsibilities. Will one audit team audit the site, or are multiple auditors responsible? How will the process links and the audit trail be managed? If one audit team is not responsible for the entire audit process, then arrangements must be made between the different auditors regarding handing off auditing processes that connect the site to its support functions. If one audit team is responsible, they should plan the sampling and the auditing of processes that link the site and its support functions whether they are at the same site or they are remote.

Support and/or Remote Locations That Are Already Assessed

Question 1: Are all outsourced processes documented, including verification of site, remote, and/or support location assessment status, if applicable?

❑ Yes ❑ No

---

* Site refers to the manufacturing location.

Support and/or Remote Locations That Are Not Assessed

Question 2: Has the organization requested an exclusion?

❑ Yes ❑ No

If the exclusion isn't justified, detail findings in the Detailed Findings Table (Figure A.8).

Step 2b: Processes Showing Sequence and Interaction, Including Key Indicators and Performance Trends, for 12 Months

The auditor should study the organization's process map or equivalent during the audit. Is the map location-specific, and does it explain the processes in the organization being audited? Many processes connect between site and remote locations. Business planning, objectives deployment, management reviews, new product development, purchasing, and sales are a few processes that have the potential of crossing functional and geographical areas. Sample the process documents. Are the interfaces of the processes clearly identified between locations, or do the documents stop within the four walls of the site or support function?

- Process maps should be simple but descriptive enough to show the sequence and interaction.
- The process map processes are meta-processes that flow into several lower-level processes typically documented in the level II procedures.
- Is there a process map that shows how all of the entities link together and how the overall processes link corporate, sales, design, manufacturing, assembly, and the warehouse?
- Does the organization's process map show the sequence and interaction of the processes at the site or entity?

Study the Linkages or Process Interfaces

The auditor must examine the links for multiple processes between the site and remote locations as well as within the entity being audited. Do the inputs and outputs match? Does the process interface make sense relative to the process being studied? In Chapter 6, three audit trails were introduced that will help auditors evaluate links, and samples that can check these interfaces.

Note: The auditor should be aware of a process approach versus an elemental, departmental, or functional approach. Processes identified by the organization should not be repeats of the clauses in AS9100 and should not be departmental or functional processes.

Process Performance

How are the processes measured and monitored? Study the process performance data to ensure the process performance.

- 3.2.14: Effectiveness—extent to which planned activities are realized and planned results achieved
- 3.2.15: Efficiency—relationship between the results achieved and the resources used

Reference: ISO 9000

Question 3: Has the organization provided a process description showing all the interactions?

❑ Yes ❑ No

Question 4: Has the organization adequately defined its processes?

❑ Yes ❑ No

If the answer is no, detail findings in the Detailed Findings Table (Figure A.8).

Question 5: Have the sampled processes been documented? If not, how have they ensured that processes are in control?

❑ Yes ❑ No

Question 6: Is process performance monitored for the processes sampled?

❑ Yes ❑ No

Study the process map overall. Ensure that it is not clause and/or function focused and that it fully explains the organization, including site and/or remote locations. Support processes could include business planning, new product development, purchasing, sales, and warehousing.

If there is a common process between the support location and the site, answer the following questions:

- Is process continuity maintained between the support location and the site?
- Are there two process owners or one?
- How is the process managed, measured, or improved?
- Is the shared process equally applicable to both the support location and the site?
- Do the outputs of the support process input directly to the site process and vice versa?
- If there are two processes, one for the support location and one for the site, are the inputs and outputs of each process clearly defined?

Note: The auditor must obtain a copy of the process description that shows all the interactions and attach it to the audit report.

The process description must show the interaction and sequencing of the ogranization's processes. The following is a partial list of possible processes.		
Customer oriented processes (COPs)	**Support oriented processes**	**Management oriented processes**
Market analysis	Maintenance	Business planning
Bid/tender	Training	Management/business review
Order/request	Human resources	Internal auditing
Product/process verification/validation	Purchasing	Continual improvement
Manufacturing	Calibration	Analysis of data
Delivery	Laboratories	Customer satisfaction
Payment		Customer requirements gathering
Warranty service		Contract review
Post-sale/customer feedback		

Figure A.3 Customer oriented processes.

If the answer to question 6 is *no,* detail findings in the Detailed Findings Table (Figure A.8) and report the determination of the stage 1 audit as not recommended to proceed to a stage 2 audit.

Study the links between processes and identify samples for the stage 2 audit. If the links are suspect, identify them either in the Process Audit Worksheet/PEAR or in the assessment planning table.

Step 2c: Conduct Document Review and Confirm That the Processes Address All Requirements of AS9100/AS9110/AS9120

Study the customer oriented processes (COPs) and the support oriented processes (SOPs). What management processes enable the COPs and SOPs? Does the process list take into consideration all the clauses and subclauses of AS9100? The audit checklist in Appendix C of this book provides a list (matrix) of subclauses and their associated processes. This list, or matrix, can be completed by the auditee prior to the audit and then verified by the auditor. *Note*: This is a high-level check, and the auditor will look in detail at processes during the document review.

Process Documentation

As part of step 2c, it's necessary to study the process documentation typically thought of as level II documentation. The auditor must ensure that it describes processes that address both AS9100/AS9110/AS9120 and the customer quality management system–specific requirements.

- Study each "shall" of AS9100/AS9110/AS9120 and the organization's customer-specific requirement document. Are these requirements addressed in the process documentation? In other words, does it give a sense of whether they are accomplishing or implementing a requirement (that is, "shall"). *Note*: Auditors should not accept process documentation that's simply AS9100/AS9110/AS9120 quoted word for word.

- If the documentation does not explain how the organization addresses a "shall," ask the auditee to show how he or she addresses it.

- The level II processes must describe *what* the organization does, *who* does it, and *when* it's done. The "what" must describe the process steps.

Note: A process is *not* a clause or element of AS9100. A process will probably not be confined within a single organizational department. In fact, it may not even be confined within the building!

When reviewing a process document, look for the following:

- Process characteristics (for example, process owners, process definition, documentation, measuring, monitoring, and continual improvement).
- Are customer oriented processes clearly documented, with clear inputs from the customer and outputs going back to the customer?
- Has the organization identified the management and support processes?
- When studying supplier management, product realization, or other processes that include outsourced processes, ensure that the organization has exercised control over these processes.
- Update the Documentation and Process Cross-Reference for AS9100C (Figure A.9).

Question 7: Are processes measured and monitored? How?

❑ Yes ❑ No

Question 8: Do the processes within the QMS address all the "shalls" of AS9100 as well as customer-specific quality management system requirements?

❑ Yes ❑ No

If the answer is no, detail findings in the Detailed Findings Table (Figure A.8)

Question 9: Were samples taken from the documentation and process cross-reference matrix?

❑ Yes ❑ No

If the answer is yes, list the samples:

Step 2d: Confirm That a Quality Manual Is Available for Each Site to Be Audited

Study the quality manual. Each audited site is required to have its own quality manual as a good practice. If the quality manual is generic, evaluate the scope, process map, and the detailed processes to assess their applicability to the site. It's likely that the process map is a generic one for the entire organization, and it may not be specific or applicable to the scope of the site and its supporting functions. At a minimum, the quality manual needs to contain three things:

- Scope of the QMS, including details of any exclusions (see step 4 in Chapter 7)
- Documented procedures or references to them (see step 2c in Chapter 7)
- Description of the interaction between the processes of the quality management system (see step 2b in Chapter 7)

Question 10: Is a quality manual available for each site to be audited?

❑ Yes ❑ No

Question 11: Does the quality manual include:

❑ Scope of the QMS, including details of any exclusions (see step 2a)

❑ Documented procedures or references to them (see step 2c)

❑ Description of the interaction between the processes of the QMS (see step 2b)

Question 12: Is there a revision number or date for each quality manual? (see step 2d)

❑ Yes ❑ No

If the answer is yes, list each revision number and/or date:

STEP 3: CONDUCT CUSTOMER FOCUS AND PERFORMANCE ANALYSIS

Step 3a: Study Customer Performance: Scorecard Results, Policy, Performance Objectives and Targets, and Past Customer Failures

AS9100 performance expectations include customer satisfaction (perception), customer satisfaction supplemental (scorecards), customer complaints and problem solving, and overall performance. All of these present potential risks to the customer or the organization. An AS9100 audit begins with the auditor analyzing overall customer satisfaction and organizational performance based on the previous criteria. The results are then linked by the auditor to poorly performing or suspect processes. These are documented in the Process Audit Worksheet/PEAR. The auditor also takes these results into consideration when auditing clause 5.2, Customer Focus, and investigates how the organization responds when performance falls short. The auditor then follows that audit trail to objectives setting, objectives deployment, and business planning. See Chapter 6 for more information on audit trails.

Having no problems in customer metrics or customer scorecards does not necessarily mean high satisfaction. Bottom line: satisfaction can only be gauged by getting the customer's perceptions. This entails asking the customer—or gauging from the customer's perceptions an answer to—some variant of the question, "Overall, how satisfied are you with us?"

Customer Scorecards

Poor customer scorecards and poor performance relative to customer expectations are indicators of potentially severe customer dissatisfaction. The auditor is required to study performance relative to all subscribing customers. If they have an online scorecard, the auditor is required to study the most current data and ascertain the supplier score.

Be careful when reviewing these data, since each customer has a different rating system. Many times, the customer is happy with one commodity or one set of parts and not another commodity or another set of parts. It is easy to show the auditor just the good data and bamboozle him/her if he/she does not really understand the products and/or the customer's method for reckoning scorecards. It is important that the auditor know all the products supplied to a particular customer and ensure that the data relative to all the products get reviewed online, live on the customer website.

The auditor should make a record of the quality issues, warranty issues, and delivery problems identified in the scorecards, product by product. Each of these issues must be tracked down and audited.

Customer Complaints and Problem-Solving Efforts

The auditor should look for patterns of the same problem repeating among common parts—and/or many seemingly random returns—which may indicate a lack of process control for a single product.

The auditor should study how the organization prioritizes issues. Does it analyze Pareto charts? Are they analyzing data by part groups or process groups? If the organization does not analyze the data meaningfully for problem solving, that may be grounds for a nonconformity or an opportunity for improvement. Having many problems or problems that continue unchecked is also an indicator of poor problem-solving capabilities or process control issues. Find out how the organization calculates ppm levels. The auditor should analyze the customer complaint database to sample issues for an audit trail. Continual problems with one issue or part number, or many different problems for one part, part family, or department, suggest potential samples for the auditor.

The key to customer-related problems and issues is the analysis of customer-related data, with prompt reaction to customer issues. When assessing the readiness of the organization, the auditor should not only evaluate customer-related issues, but also how the organization responds to them.

Identifying Risks to the Customer

Overall, customer satisfaction issues, customer scorecard issues, and customer complaints about product quality all constitute risk to the customer. These issues should all be investigated during the on-site audit. A second set of internal criteria may also prove to be a risk to the customer. Internal audit performance issues and overall performance issues (see step 3b) constitute risks to the customer as well. Internal issues will often translate to customer issues. Step 3c will show how the auditor can translate these issues to on-site process investigations in the stage 2 audit.

The auditor shall review the following, as a minimum, for each customer:

- Identify all organizational key customers.
- From the list of customers, identify which customers provide scorecards. For each customer, review the customer scorecard online. Carefully identify which customers score by plant and which customers score by product line. Don't be satisfied by reviewing printed-out materials because they could be misleading.

- Identify all customer performance issues. In particular, determine whether the customer has put them into a special status category for improvement.
- For each customer, ask the organization how it evaluates "overall satisfaction." (*Note:* Scorecard performance isn't the same as customer satisfaction.)

For any customer scorecard category where the organization's performance doesn't meet the customer expectation and/or goal, document the category and related suspect processes in the Stage 2 Assessment Plan (Figure A.7).

Question 13: Have all customers, customer scorecards, and customer performance issues been identified?

❑ Yes ❑ No

Question 14: Is there an overall process for gathering data about customer needs and expectations? This isn't a contract review process. Does the process show all interactions with the customer, how data are used for taking action to satisfy the customer, and how the organization sets customer-focused objectives?

❑ Yes ❑ No

If the answer is no, detail findings in the Detailed Findings Table (Figure A.8)

Question 15: Is there a prioritized list of customer expectations?

❑ Yes ❑ No

Question 16: Does the list include cost, quality, delivery, and other key customer issues?

❑ Yes ❑ No

Question 17: Are objectives set based on the customer expectations gathered?

❑ Yes ❑ No

Note: To help determine the answer to question 17, complete the Customer Expectation Sampling Sheet (Figure A.5). See also the management systems audit trail in Chapter 6 and keep in mind the following:

- Is there a process to gather and analyze customer satisfaction (perception) and the customer scorecards?

Information from Customer Scorecard				
Customer	Customer quality performance	Customer/ assembly plant disruptions	Delivery schedule performance	Other

Figure A.4 Information from Customer Scorecard.

Customer Expectation Sampling Sheet					
Customer expectations*	Objectives	Deployed objectives/ departments	Related processes	Plan to meet objectives	Evidence of objectives being met
* Is auditor sampling product quality, delivery, technology, lead time, reliability, cost, and so on? *Note:* Which dimensions are important to the customer?					

Figure A.5 Customer Expectation Sampling Sheet.

- From the list of customers, identify which customers provide scorecards. For each customer, review the customer scorecard online. Carefully identify which customers score by plant and which customers score by product line. Don't be satisfied by reviewing printed-out materials because they could be misleading.
- Identify all customer performance issues. In particular, determine whether the customer has put them into a special status category for improvement.
- For each customer, ask the organization how it evaluates "overall satisfaction." (*Note:* Scorecard performance isn't the same as customer satisfaction.)

Question 18: When the organization fails to meet any customer expectation, is the problem addressed through root cause analysis followed by systemic corrective action?

❑ Yes ❑ No

If the answer is no, detail findings in the Detailed Findings Table (Figure A.8)

Question 19: Are there any open customer complaints?

❑ Yes ❑ No

Question 20: If yes, has a corrective action plan been implemented or proposed? (*Note:* Corrective action implemented or proposed should include root cause analysis followed by systemic corrective action.)

❑ Yes ❑ No

If the answer is no, detail findings in the Detailed Findings Table (Figure A.8)

Review and analyze the last 12 months of customer complaint history, and document any trends, along with the related suspect process, in the Assessment Planning Table (Figure A.6). Review the data for the following trends:

- Repeat issues

Assessment Planning Table	
Customer and performance issues	Related suspect processes

Figure A.6 Assessment Planning Table.

- Serious customer issues (for example, prism, yard hold, spills, or recalls)
- Blip or trend in performance

Review the organization's Pareto analysis for trends in departments, product families, manufacturing processes, or design issues.

Step 3b: Evaluate Internal Audit and Management Review Results from the Previous 12 Months

Internal Audits

Internal audits are a good gauge of how well the organization understands itself. The auditor is reviewing the internal audit to ensure that the organization has conducted a complete system audit that includes all the processes and all the clauses of AS9100/AS9110/AS9120. The organization is required to have 12 months of audit history.

Audits should be scheduled based on status, importance, and the organization's annual plan. Also, the audits must be based on customer complaints, internal/external performance data, and how the internal audit has considered the customer-specific quality management system requirements.

Study the quality of the audit and the nonconformities issued. Does the internal audit include all the issues noticed in the organization thus far? Is the audit adequate? The nonconformities issued should have three parts: nonconformity, quote of the requirement, and the objective evidence. What is the quality of the nonconformities? Are they clear and concise?

Check out the quality of the nonconformity closeouts. Is there objective evidence to show that the corrective action has been implemented? Also, is there evidence that the system corrective actions have been implemented? Is there evidence to show that the problem will not repeat?

System Audit

System audits are conducted periodically (minimum once a year) to provide top management an overview of the quality management system. System audits should be conducted with the same formality as third-party audits, and should use the same processes and time durations as an initial audit. System audits should cover all the process map processes and all the clauses in AS9100. *Note*: System audits are not a series of short audits conducted monthly, but are a snapshot in time of the overall health and vitality of the QMS.

The intent of these audits is to ascertain whether the overall system is "effective and efficient." This is the formal audit, which needs to be conducted similarly to an external audit. In this audit, the auditors are ensuring that the organization is moving toward its goals and objectives and that customer satisfaction, including the customer satisfaction supplemental metrics, is being met.

Process Approach versus Clause or Elemental Approach

The audit must follow the aerospace auditing approach (see Chapter 5). The audit plan must have processes from the organization's process map. Processes aren't chosen randomly, but are prioritized based on risks to the customer (for example, customer satisfaction, customer complaints, and COPs).

Overall Performance

The organization's overall performance must be gauged by examining records of management review. *Note*: Sometimes auditees only conduct a management review once a year to comply with AS9100, and then only to show the records to the auditor. This type of "compliance" for such an important requirement should be duly recognized as a major nonconformity.

The management review must be conducted at suitable intervals to assess overall improvements and to note whether the organization is meeting business objectives and satisfying its customer needs and expectations. It is important for the auditor to note whether the management review is just a presentation of facts or it is a meeting that is improvement oriented and evaluates the need for changes to the overall management system, quality policy, and objectives.

At a minimum, the management review or business review must cover these topics:

- Results of audits

- Customer feedback
- Process performance and product conformity
- Status of preventive and corrective actions
- Follow-up actions from previous management reviews
- Planned changes that could affect the QMS
- Recommendations for improvement

The auditor shouldn't expect the organization to cover each topic during every business review. However, the topics must be covered according to top management requirements to move the organization forward. Also, it is important to evaluate the management review's output to ensure that it includes decisions and actions for improvement of the quality management system, processes, product related to customer requirements, and resource issues.

Overall performance should be gauged according to the requirements of AS9100 by examining the business reviews that move the company forward on a weekly and monthly basis. Review the key indicators of the business and note those that are performing poorly. Assess the overall quality of the business reviews. Is the company progressing toward its objectives? Also, do the objectives reflect customer needs, expectations, and key concerns?

Measuring Key Indicators and Performance Trends

Although it's not a requirement, the AS9100 auditor expectation is that the organization will measure trends on a chart that shows variables on the y-axis and months on the x-axis. Omnex recommends the use of trend charts, Pareto charts, and summaries of actions taken to improve a key indicator.

Question 21: Has a complete management review cycle been conducted that includes 12 months of records? The management meeting records should cover the minimum requirements of AS9100; refer to clause 5.6.

❑ Yes ❑ No

If the answer is no, detail findings in the Detailed Findings Table (Figure A.8) and report the stage 1 determination as not recommended to proceed to a stage 2 audit.

Question 22: Is there output from the management review that's action and improvement oriented?

❑ Yes ❑ No

If the answer is no, detail findings in the Detailed Findings Table (Figure A.8).

Question 23: Does the organization track a minimum of 12 months of operational performance indicators? Performance indicators are internal indicators, such as internal ppm, yield, inventory turns, and availability, that demonstrate operational performance.

❑ Yes ❑ No

Question 24: Are corrective action plans established for indicators that don't meet established goals? For those indicators not meeting the goals, document these indicators in the Assessment Planning Table (Figure A.6) along with the related suspect processes.

❑ Yes ❑ No

If the answer is no, detail findings in the Detailed Findings Table (Figure A.8).

Question 25: Are the organizational objectives and goals consistent between the objectives and the quality policy?

❑ Yes ❑ No

If the answer is no, detail findings in the Detailed Findings Table (Figure A.8).

Question 26: Has a complete internal audit been conducted? (Review 12 months of internal audit history.) The internal audit should have been completed, covering all processes and clauses.

❑ Yes ❑ No

If the answer is no, detail findings in the Detailed Findings Table (Figure A.8).

Question 27: Are all nonconformities resulting from the full internal system audit closed? Internal audit nonconformities shall be addressed through root cause analysis and systemic corrective action.

❑ Yes ❑ No

If the answer is no, detail findings in the Detailed Findings Table (Figure A.8).

Step 3c: Identify Suspect Processes (That Is, Those Suspected of Poor Performance Based on Customer and Performance Data Analysis)

Based on the analysis of customer and performance data, identify poorly performing processes that create risk for the customer. Using the Stage 2 Assessment Plan (Figure A.7), analyze the customer supplemental, customer scorecards, management review performance data, and customer complaints. What are the key customer performance issues?

Identify the suspect processes that affect performance. Use the process identified in step 3b (see Chapter 7) to identify the relationship between results and process performance. Prioritize the processes as they relate to customer risk and organizational risk. Based on analysis of these data, document customer, supporting, and/or management oriented processes that show weakness and require increased focus during the assessment planning stage.

An audit plan should be organized according to processes from the organization's process map, not by clauses in the standard. The audit plan should be prioritized according to "suspect" processes identified during customer focus and performance analysis. Next, the auditor should identify the sequence of processes to audit by referring to the audit trails discussed in Chapter 6: business planning and management (BPM) review, new product development (NPD), and provision audit trails. Study the process map and identify the BPM, NPD, and provision audit trails.

- Audit the BPM processes and top management at the beginning of the audit.
- Start with customer expectations, customer satisfaction, and scorecards. Review top management's ideas about key issues from the customer's viewpoint.
- Identify which auditor is the most proficient in each audit trail and update the audit plan with the processes related to BPM, NPD, and provision audit trails.

Stage 2 Assessment Plan					
Objective: To verify conformance to AS9100					
Date	**Time**	**Auditor**	**Location**	**Organization's process #/description**	**Standard clauses**
				Opening meeting	
				Facility tour	
				Review of customer scorecard and associated corrective action	

Figure A.7 Stage 2 Assessment Plan.

- Identify all the organization's suspect processes and COPs to ensure they're included in the audit plan (see Chapter 6). Complete the Documentation and Process Cross-Reference for AS9100C (Figure A.9) and ensure that all clauses are being audited.
- Identify OEM and other customer-specific quality management system requirements and make notations in the audit plan to ensure that customer-specific requirements (CSRs) will be sampled.
- Audit based on the defined processes of the organization and not the AS9100 clauses.
- Begin with an audit of top management and cover the following:
  - Areas of risk to the customer, including customer complaints and special status notification

- Internal audits
- Management review and actions
- Progress toward continual improvement as related to set objectives
- Effectiveness of corrective actions
- Verification of corrective action taken since previous audit

Step 3d: Confirm the Customer-Specific Quality Management System Requirements to Be Included in the Audit

Each on-site audit (that is, initial, surveillance, and recertification) shall include an audit of the:

- Organization's implementation of new customer requirements since the last audit
- Customer complaints and the organization's responses
- Organization's internal audit and management review results and actions
- Progress made toward continual improvement targets
- Effectiveness of corrective actions and verification since the last audit
- Effectiveness of the QMS in achieving both customer and organization objectives

Every audit shall include auditing on all shifts. Manufacturing activities shall be audited on all shifts, where they occur.

Question 28: Has the organization adequately defined its customers and the supplementary QMS requirements?

❑ Yes ❑ No

If the answer is no, detail findings in the Detailed Findings Table (Figure A.8).

Question 29: Does the organization have a process to gather and update customer-specific quality management system requirements?

❑ Yes ❑ No

If the answer is no, detail findings in the Detailed Findings Table (Figure A.8).

STEP 4: DETERMINE THE APPROPRIATE SCOPE

The scope is inherently linked to the process approach, and especially the "site" and the "remote location functions." Customers and customer-specific quality management system requirements also affect the scope.

Question 30: Does the scope adequately reflect the organization's operation? When determining the scope, the requirements in AS9100 clause 1.2 must be taken into account.

❑ Yes ❑ No

If the answer is no, detail findings in the Detailed Findings Table (Figure A.8).

STEP 5: DETERMINE THE ORGANIZATION'S READINESS

In third-party audits, the readiness of the organization to proceed to stage 2 is determined. If the organization isn't ready to conduct the stage 2 site audit, the certification body and the organization can agree to stop the process.

In the internal audit, this is really not a choice. If the organization is not ready, the internal auditors can issue nonconformances that can be added to the final audit report or closed out during the stage 2 audit (see Figure A.8).

Stage 1 audit results shall be documented and communicated to the organization.

The stage 1 audit is the most critical step of the audit, so auditors should be thorough and provide nonconformities as necessary for improvement.

For guidance to internal auditors, the following situations typically may require postponement in a third-party situation, or a major nonconformity in an internal audit situation:

Detailed Findings Table	
Nonconformities identified during this stage 1 assessment	
NC#	**Nonconformity description**

Figure A.8 Detailed Findings Table.

- Customer scorecards show that the organization is under a special status category. Some of the customer-specific organizational approval statuses include limited approval, probation, suspension, or withdrawal.
- The organization doesn't have one year of internal audits, management reviews, or performance data.
- An internal system audit to AS9100/AS9110/AS9120 (all processes, clauses, or aerospace process approach to audits) hasn't been completed.
- Management review shows no top management involvement, or the management review is incomplete.
- The organization shows poor customer or process focus.
- Documentation shows many requirements not being addressed by the organization's processes.

If there are obvious major nonconformities with respect to the implementation of the management system, and/or performance or customer issues, the auditor notes them and issues major nonconformities, or identifies them in the audit trails for the stage 2 audit.

STEP 6: PREPARE AND DELIVER THE STAGE 1 REPORT

Internal and supplier auditors can identify any major or minor nonconformities and report them in the internal audit.

STEP 7: CREATE AN AUDIT PLAN

Creating a prioritized audit plan based on weaknesses in customer and organizational performance is a key to the aerospace auditing process approach. Following the audit trails described in Chapter 6 is a good method for understanding linkages between processes and the AS9100 clauses, and for sample taking.

Study the organization's process map and identify the three audit trails—business planning and management (BPM), new product development (NPD), and provision audit trails (see Chapter 6).

- Audit the BPM processes and top management in the beginning of the audit.
- Start with customer expectations, customer satisfaction, and the scorecards. Obtain top management's thoughts and ideas on key issues from the customer's viewpoint.
- Identify which auditor is more proficient in each audit trail, and update the audit plan with the processes related to BPM, NPD, and provision audit trails.
- Identify all suspect processes and COPs in the organization and ensure they are in the audit plan.
- Complete the process-to-clause checklist and ensure that all clauses are being audited.
- Identify OEM and other customer-specific quality management system requirements, and make notations in the audit plan to ensure that the CSRs will be sampled.

When creating the audit plan, ensure that the following items are available for the audit:

- Description of processes showing their sequence and interactions, including key indicators and performance trends for the previous 12 months, minimum
- Evidence that the organization's processes address all the requirements of AS9100
- Quality manual for each site to be audited
- Internal audit and management review planning and results from previous 12 months
- List of customer-specific quality management system requirements
- Customer satisfaction and complaints status, including customer reports and scorecards

Finally, the auditor should keep in mind:

- Analysis of actual or potential risk to the customer, product, and processes
- Demonstration of links between audit trails
- Auditing manufacturing activities on all shifts, where they occur
- Optimizing audit time based on the organization's layout
- Interfaces with the remote support functions
- Customer-specific quality management system requirements and all relevant processes, including new customers since the last audit
- Customer concerns and/or complaints, special status notification, and the organization's response
- Internal audit and management review results and actions
- Progress made toward continual improvement
- Effectiveness of the corrective actions and verification since the last audit
- QMS effectiveness with regard to achieving both customer and organizational objectives
- Distributing the audit plan to the organization and all audit team members

Documentation and Process Cross-Reference for AS9100C				
Clause	Clause heading	Organization's process number/ description/document reference(1, 2)	Process owner[1]	Stage 1 results[3] (C–O–N/A)
4.1	(Quality management system) general requirements			
4.2	Documentation requirements			
4.2.1	(Documentation requirements) general			
4.2.2	Quality manual			
4.2.3	Control of documents			
4.2.4	Control of records			
5.1	Management commitment			
5.2	Customer focus			
5.3	Quality policy			
5.4.1	Quality objectives			
5.4.2	Quality management system planning			

Note 1: Shaded areas in light gray are to be completed by the organization (auditee); areas shaded in dark gray are to be completed by the auditor

Note 2: Process numbers/descriptions/document reference must be linked to the organization's (auditee's) process map

Note 3: C refers to "compliance," O refers to "observations," and N/A is "not applicable." N/A is only acceptable for product design.

Figure A.9 Documentation and Process Cross-Reference for AS9100C.

Continued

Clause	Clause heading	Organization's process number/ description/document reference[1, 2]	Process owner[1]	Stage 1 results[3] (C–O–N/A)
5.5.1	Responsibility and authority			
5.5.2	Management representative			
5.5.3	Internal communication			
5.6.1	(Management review) general			
5.6.2	Review input			
5.6.3	Review output			
6.1	Provision of resources			
6.2.1	(Human resources) general			
6.2.2	Competence, training and awareness			
6.3	Infrastructure			
6.4	Work environment			
7.1	Planning of product realization			
7.1.1	Project management			
7.1.2	Risk management			
7.1.3	Configuration management			
7.1.4	Control of work transfers			
7.2	Customer-related processes			

Figure A.9 *Continued.*

Clause	Clause heading	Organization's process number/ description/document reference[1, 2]	Process owner[1]	Stage 1 results[3] (C–O–N/A)
7.2.1	Determination of requirements related to the product			
7.2.2	Review of requirements related to the product			
7.2.3	Customer communication			
7.3	Design and development			
7.3.1	Design and development planning			
7.3.2	Design and development inputs			
7.3.3	Design and development outputs			
7.3.4	Design and development review			
7.3.5	Design and development verification			
7.3.6	Design and development validation			
7.3.6.1	Design and development verification and validation testing			
7.3.6.2	Design and development verification and validation documentation			
7.3.7	Control of design and development changes			
7.4	Purchasing			
7.4.1	Purchasing process			

Figure A.9 *Continued.*

Clause	Clause heading	Organization's process number/ description/document reference[1, 2]	Process owner[1]	Stage 1 results[3] (C–O–N/A)
7.4.2	Purchasing information			
7.4.3	Verification of purchased product			
7.5	Production and service provision			
7.5.1	Control of production and service provision			
7.5.1.1	Production process verification			
7.5.1.2	Control of production process changes			
7.5.1.3	Control of production equipment, tools and software programs			
7.5.1.4	Post-delivery support			
7.5.2	Validation of processes for production and service provision			
7.5.3	Identification and traceability			
7.5.4	Customer property			
7.5.5	Preservation of product			
7.6	Control of monitoring and measuring equipment			
8.1	(Measurement, analysis and improvement) general			

Figure A.9 *Continued.*

Clause	Clause heading	Organization's process number/ description/document reference[1, 2]	Process owner[1]	Stage 1 results[3] (C–O–N/A)
8.2	Monitoring and measurement			
8.2.1	Customer satisfaction			
8.2.2	Internal audit			
8.2.3	Monitoring and measurement of processes			
8.2.4	Monitoring and measurement of product			
8.3	Control of nonconforming product			
8.4	Analysis of data			
8.5	Improvement			
8.5.1	Continual improvement			
8.5.2	Corrective action			
8.5.3	Preventive action			

Figure A.9 *Continued.*

STEP 8: CONDUCT PROCESS ANALYSIS AND PREPARE A CUSTOMIZED PROCESS WORKSHEET

Preparing the Process Audit Worksheet/PEAR

The auditor should use the Process Audit Worksheet/PEAR in Figure A.10 for auditing all processes. The process analysis allows the auditor to evaluate inputs, outputs, resources, measurement, monitoring, and methods. The performance analysis and process analysis should assist the auditor in the readiness review in developing process-related questions in the process worksheet.

Process Audit Worksheet/PEAR

Company name:	**Location:**	**Audit type:**	**Standard:**
Process characteristics: ❑ Has responsibilities/process owner assigned? ❑ Considers customer specifics ❑ Is documented? ❑ Is monitored?		**Classification:** ❑ Needs further research ❑ Nonconformities ❑ Opportunities for improvements	
Auditor:		**Process/area:**	
Organization's method for determing process effectiveness (for example, key indicators or measurements):		**Applicable clauses (AS9100/9110/9120):**	
Auditees/interviewees: **Audit observations and comments supporting process effectiveness determination:**		**Process interfaces (process linkages) and process details:** **Process activities:**	
Statement of effectiveness:	1. ❑ Not implemented; planned results are not achieved. 2. ❑ Implemented; planned results are not achieved and appropriate actions not taken. 3. ❑ Implemented; planned results are not achieved, but appropriate actions are being taken. 4. ❑ Implemented; planned results are achieved.		

Interactions

Description of audit observations, evidence, potential and actual findings	**Description of audit observations, evidence, potential and actual findings**
Customer issue or process weakness noted from stage 1: Process customer: Key indicator description: Key indicator actual performance: Description of corrective action if performance target isn't met: **Process design and robustness:** (Does process design allow performance objectives? Does it meet AS9100/9110/9120 requirements?) **Customer-specific quality management system requirements, including statutory and regulatory, affected (which customer?):**	**Needed resources:** What? (Materials/equipment) Who? (Competence/skills/training) Inputs → PROCESS → Outputs How? (Methods/procedures/techniques) Criteria (Measurement/assessment)

Figure A.10 Process Audit Worksheet/PEAR example.

Clause		Process conformity										NCR number
		1	2	3	4	5	6	7	8	9	10	
4.1	(Quality management system) general requirements											
4.2	Documentation requirements											
4.2.1	(Documentation requirements) general											
4.2.2	Quality manual											
4.2.3	Control of documents											
4.2.4	Control of records											
5.1	Management commitment											
5.2	Customer focus											
5.3	Quality policy											
5.4.1	Quality objectives											
5.4.2	Quality management system planning											
5.5.1	Responsibility and authority											
5.5.2	Management representative											
5.5.3	Internal communication											
5.6.1	(Management review) general											
5.6.2	Review input											

Figure A.11 Process matrix.

Clause		Process conformity										NCR number
		1	2	3	4	5	6	7	8	9	10	
5.6.3	Review output											
6.1	Provision of resources											
6.2.1	(Human resources) general											
6.2.2	Competence, training and awareness											
6.3	Infrastructure											
6.4	Work environment											
7.1	Planning of product realization											
7.1.1	Project management											
7.1.2	Risk management											
7.1.3	Configuration management											
7.1.4	Control of work transfers											
7.2.1	Determination of requirements related to the product											
7.2.2	Review of requirements related to the product											
7.2.3	Customer communication											
7.3.1	Design and development planning											
7.3.2	Design and development inputs											
7.3.3	Design and development outputs											

Figure A.11 *Continued.*

Clause		Process conformity										NCR number
		1	2	3	4	5	6	7	8	9	10	
7.3.4	Design and development review											
7.3.5	Design and development verification											
7.3.6	Design and development validation											
7.3.6.1	Design and development verification and validation testing											
7.3.6.2	Design and development verification and validation documentation											
7.3.7	Control of design and development changes											
7.4.1	Purchasing process											
7.4.2	Purchasing information											
7.4.3	Verification of purchased product											
7.5.1	Control of production and service provision											
7.5.1.1	Production process verification											
7.5.1.2	Control of production process changes											
7.5.1.3	Control of production equipment, tools and software programs											
7.5.1.4	Post-delivery support											
7.5.2	Validation of processes for production and service provision											
7.5.3	Identification and traceability											

Figure A.11 *Continued.*

Clause		Process conformity										NCR number
		1	2	3	4	5	6	7	8	9	10	
7.5.4	Customer property											
7.5.5	Preservation of product											
7.6	Control of monitoring and measuring equipment											
8.1	(Measurement, analysis and improvement) general											
8.2.1	Customer satisfaction											
8.2.2	Internal audit											
8.2.3	Monitoring and measurement of processes											
8.2.4	Monitoring and measurement of product											
8.3	Control of nonconforming product											
8.4	Analysis of data											
8.5	Improvement											
8.5.1	Continual improvement											
8.5.2	Corrective action											
8.5.3	Preventive action											

Figure A.11 *Continued.*

Process/activities/subjects	Comments
Process sequence and interaction (for example, process maps, flowcharts)	
High-risk processes/products	
Risk management	
Special processes (for example, metal joining, coating, thermal processing, bonding, chemical treatment)	
Regulatory requirements/authority approval/recognitions	
Configuration management	
Project/program management	
Continual improvement activities	
Special requirements/critical items (including key characteristics)	
First article inspection (for example, AS9102)	
Foreign object debris/damage (FOD) programs	
Special work environment (for example, electrostatic discharge sensitive [ESDS], clean room, temperature/humidity controls)	
Customer presence in organization (for example, on-site representative, regular meetings, reason)	
Restricted areas/proprietary information/confidentiality	
Export limitations/controls	
Customer-delegated inspection	
Nonconforming product management (for example, delegated materials review board [MRB])	
Customer satisfaction and complaints status	
Customer authorized direct ship/delivery	

Figure A.12 Stage 1 completion checklist.

Appendix B
Confidential Assessment Report for Stage 2

CONTENTS

(See Chapter 8 for more details)

Organization name:			
City, State:			
Organization contact:			
Lead auditor:			
Audit scope:			
Permitted exclusions:			
Nonconformity			
Total number of nonconformities (issued during audit):			
Major		Minor	
PEAR			
Total number of PEARs (issued during audit):			
Level 1	Level 2	Level 3	Level 4

Figure B.1 Stage 2 audit report.

ORGANIZATION INFORMATION

Organization name: ______________________________

Contact person: ______________________________

Department: ______________________________

Telephone: ______________________________

Fax: ______________________________

E-mail: ______________________________

Street: ______________________________

City: ______________________________

State or province: ______________________________

Zip or postal code: ______________________________

Standard: AS9100C

Type of assessment: Stage 2

Assessment team: ______________________________

Person days: ______________________________

Start date: ______________________________

End date: ______________________________

Assessment information: ______________________________

____

ASSESSMENT REPORT ACCEPTANCE

Signed for on behalf of auditor:

Name: ______________________________

Date: ______________________________

Signed for on behalf of organization:

Name: ______________________________

Date: ______________________________

STEP 9: CONDUCT AUDIT OF REMOTE SUPPORTING FUNCTIONS

(See Figure C.1)

Site and/or remote and/or support locations.* Please detail individually all sites, including remote locations that support this site (include location name, address, and processes performed that support the site). A support location may be remote to the site or embedded within one site and supporting numerous sites.

> *Example 1.* Site and corporate headquarters located at same location. Corporate headquarters includes human resources, sales, purchasing, and other functions that support other sites.

---

* Site refers to the manufacturing location.

Audit Conclusions

Audit summary:
Key issues/concerns requiring top management attention:
Strengths and good practices:

Figure B.2 Audit Conclusions.

Opportunities for improvement/observations:

Previous audit nonconformity status:		
NCRs issued (during last audit)	NCRs closed:	NCRs open:

Changes to organization/facilities/quality management system/scope (since last visit):			
		(As applicable)	
Ref. number	Brief description	Organization document ref.	9100/9110/9120 clause ref.

Strengths and good practices:

Figure B.3 Opportunities for improvement/observations.

Example 2. Site A includes a product design center, which supports numerous sites.

Omnex recommends that the site and the supporting functions be audited together in sequence, starting with the supporting functions.

Organization and/or Locations Information Approval Checklist

- Identify processes that link between the support functions and the site. These processes should be evident in the process map itself. Are they?

  ❑ Yes ❑ No

- Take samples for each of the processes that will be used to test the interface. Whether these are top management objectives, new products in sales and/or design, or customer satisfaction information input into the sales office, carefully sample them and follow each process from the supporting function into the site. Does the process link?

  ❑ Yes ❑ No

- Does the process documentation connect the site to the remote location?

  ❑ Yes ❑ No

- How is the process managed? Is it working effectively?

  ❑ Yes ❑ No

- Is it measured, monitored, and improved?

  ❑ Yes ❑ No

Audit the process using the stage 2 steps:

- Are the supporting functions and site working together?

  ❑ Yes ❑ No

- Are they designed to succeed together?

  ❑ Yes ❑ No

Employee Shift Details—Site/Remote/Support Location							
No.	Location name	Aerospace/ other	Street	City	Postal code/ zip	Country	Already assessed to AS9100C?

Figure B.4 Employee Shift Details—Site/Remote/Support Location.

STEP 10: OPENING MEETING

The client will maintain the original meeting attendance list and the original for each audit locally.

- Introduce audit team and attendees.
- Pass out the attendance sheet.
- Explain the aerospace process approach.
- Review objectives, scope, and criteria:
  - Review supporting functions and interfacing processes.
- Summary of methods and procedures used for audit:
  - Auditor takes notes of details for both conformities and nonconformities.
  - Audit conclusion is based on samples taken:
    - The audit will take samples from a cross-section of the organization, from top management to maintenance workers and engineers.
  - The audit is restricted to small groups of three or fewer people.
  - The auditor notifies auditee of nonconformities during the audit as well as during a daily review meeting.
  - Questions should be directed toward lead auditor.
  - Conditions are specified for when a major nonconformity affecting the customer is uncovered.
- Establish auditee communication link.
- Reconfirm the following:
  - Current customers
  - Customer satisfaction and complaints status, including customer reports and scorecards
  - Any customer special status (bad supplier status)
  - Supporting functions and interfacing processes
- Review prioritized audit plan.

- Confirm status of stage 1 issues, including documentation.
- Confirm time and date of closing meeting.
- Confirm relevant safety, emergency, and security procedures.

STEP 11: CONDUCTING THE AUDIT

11a. Conduct Facility Tour, If Needed

- Modify the audit plan based on information collected during the opening meeting and facility tour.
- Note on the process worksheets issues or areas to investigate.

11b. Study Customer and Organizational Performance

- Reexamine the customer scorecard. Has performance maintained the same level, or has performance deteriorated?
- Study the last business performance review. Has performance been sustained at the same level? Did the organization act and follow up on the actions identified from the management review provided during the readiness review?
- Has the organization been put on a special status notification?
- Determine whether to adjust or reprioritize the audit plan based on the latest customer and performance issues.

11c. Meeting with Top Management

Interview top managers to understand and evaluate the following:

- Establishment and continued relevance of the organization's quality policy and objectives
- Establishment of performance measures aligned to quality objectives
- Quality management system development, implementation, and continual improvement
- Top management commitment

- Quality management system performance and effectiveness
- Performance to customer expectations (for example, supplier rating, scorecard, audit results)
- Actions taken to address areas that are not meeting customer performance expectations

11d. Audit Organizational Processes

- Demonstrate the use of the aerospace process approach, customer focus, and leadership, including the prioritized audit plan:
  - Prioritize the audit plan based on risks to customer (see Chapter 7).
  - Use an AS9101D and OER checklist (see Appendix C on CD-ROM) and Process Audit Worksheet/PEAR (see Figure B.5) as a primary tool for auditing processes.
  - Customize the checklist with issues that affect the customer, customer-specific quality management system requirements, and/or items that appear to be poor process definition issues.
- Use the audit plan and the organization's defined processes, including sequence and interaction. Don't be requirements oriented; be process oriented.
- Be able to determine whether the organization is operating according to its process definition (that is, a process map and its sequence and interactions. See Chapter 7). Do the processes reflect reality?
- Audit processes to determine if each one is capable of meeting the key process indicators and is performing satisfactorily.
- Make sure that the customer-specific quality management system requirements are identified, addressed, and maintained in the business management system. (*Note:* Customer-specific requirements should be integrated into the processes.)
- Conduct interviews about the process with those that are involved in the process at its location. Avoid conference room audits. Always take samples or have the auditees show objective evidence when interviewing them to confirm their statements. The auditor should always choose the samples so as to avoid biased samples being given.

Process Audit Worksheet/PEAR

Company name:	Location:	Audit type:	Standard:
Process characteristics: ❑ Has responsibilities/process owner assigned? ❑ Considers customer specifics ❑ Is documented? ❑ Is monitored?		**Classification:** ❑ Needs further research ❑ Nonconformities ❑ Opportunities for improvements	
Auditor:		**Process/area:**	
Organization's method for determing process effectiveness (for example, key indicators or measurements):		**Applicable clauses (AS9100/9110/9120):**	
Auditees/interviewees: **Audit observations and comments supporting process effectiveness determination:**		**Process interfaces (process linkages) and process details:** **Process activities:**	
Statement of effectiveness:	1. ❑ Not implemented; planned results are not achieved. 2. ❑ Implemented; planned results are not achieved and appropriate actions not taken. 3. ❑ Implemented; planned results are not achieved, but appropriate actions are being taken. 4. ❑ Implemented; planned results are achieved.		

Interactions

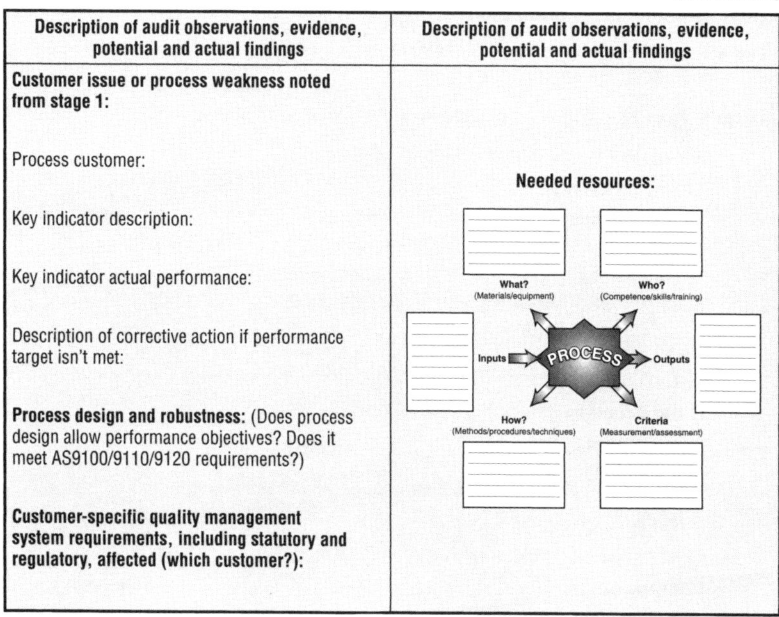

Description of audit observations, evidence, potential and actual findings	Description of audit observations, evidence, potential and actual findings
Customer issue or process weakness noted from stage 1: Process customer: Key indicator description: Key indicator actual performance: Description of corrective action if performance target isn't met: **Process design and robustness:** (Does process design allow performance objectives? Does it meet AS9100/9110/9120 requirements?) **Customer-specific quality management system requirements, including statutory and regulatory, affected (which customer?):**	**Needed resources:**

Figure B.5 Process Audit Worksheet/PEAR example.

- Document both conformity and nonconformity in the process worksheet. The information should be clear enough for an independent review by a third party, if necessary.

11e: Verify That All Processes and Clauses Are Audited

The auditor should follow the audit plan (see Figure B.6) and ensure that the audit encompassed all processes and clauses. Complete the Stage 2 Process/Clause Matrix (Figure B.7).

STEP 12: WRITE UP NONCONFORMITIES

Please respond by using your own corrective action method (for example, 7D, 8D, or five whys) in the same way that your organization would respond to a customer issue. Include the root cause analysis and systemic corrective

Assessment Plan—Stage 2					
Objective: To verify conformance to AS9100/AS9110/AS9120					
Date	**Time**	**Auditor**	**Location**	**Organization's process number and/or description**	**Clause**

Figure B.6 Assessment Plan—Stage 2.

Clause		Process conformity										NCR number
		1	2	3	4	5	6	7	8	9	10	
4.1	(Quality management system) general requirements											
4.2	Documentation requirements											
4.2.1	(Documentation requirements) general											
4.2.2	Quality manual											
4.2.3	Control of documents											
4.2.4	Control of records											
5.1	Management commitment											
5.2	Customer focus											
5.3	Quality policy											
5.4.1	Quality objectives											
5.4.2	Quality management system planning											
5.5.1	Responsibility and authority											
5.5.2	Management representative											
5.5.3	Internal communication											
5.6.1	(Management review) general											
5.6.2	Review input											

Figure B.7 Stage 2 Process/Clause Matrix.

Clause		Process conformity										NCR number
		1	2	3	4	5	6	7	8	9	10	
5.6.3	Review output											
6.1	Provision of resources											
6.2.1	(Human resources) general											
6.2.2	Competence, training and awareness											
6.3	Infrastructure											
6.4	Work environment											
7.1	Planning of product realization											
7.1.1	Project management											
7.1.2	Risk management											
7.1.3	Configuration management											
7.1.4	Control of work transfers											
7.2.1	Determination of requirements related to the product											
7.2.2	Review of requirements related to the product											
7.2.3	Customer communication											
7.3.1	Design and development planning											
7.3.2	Design and development inputs											
7.3.3	Design and development outputs											

Figure B.7 *Continued.*

Clause		Process conformity										NCR number
		1	2	3	4	5	6	7	8	9	10	
7.3.4	Design and development review											
7.3.5	Design and development verification											
7.3.6	Design and development validation											
7.3.6.1	Design and development verification and validation testing											
7.3.6.2	Design and development verification and validation documentation											
7.3.7	Control of design and development changes											
7.4.1	Purchasing process											
7.4.2	Purchasing information											
7.4.3	Verification of purchased product											
7.5.1	Control of production and service provision											
7.5.1.1	Production process verification											
7.5.1.2	Control of production process changes											
7.5.1.3	Control of production equipment, tools and software programs											
7.5.1.4	Post-delivery support											
7.5.2	Validation of processes for production and service provision											
7.5.3	Identification and traceability											

Figure B.7 *Continued.*

Clause		Process conformity										NCR number
		1	2	3	4	5	6	7	8	9	10	
7.5.4	Customer property											
7.5.5	Preservation of product											
7.6	Control of monitoring and measuring equipment											
8.1	(Measurement, analysis and improvement) general											
8.2.1	Customer satisfaction											
8.2.2	Internal audit											
8.2.3	Monitoring and measurement of processes											
8.2.4	Monitoring and measurement of product											
8.3	Control of nonconforming product											
8.4	Analysis of data											
8.5	Improvement											
8.5.1	Continual improvement											
8.5.2	Corrective action											
8.5.3	Preventive action											

Figure B.7 *Continued.*

action; failure to include these will result in your responses being rejected by the lead auditor.

Writing Nonconformities

- Nonconformity statements must contain a statement of nonconformity, the standard requirement, and objective evidence. *Note:* One nonconformity can be written to cover more than one "shall."
- Categorize nonconformities as major or minor.
- Nonconformities should be cross-referenced to the organization's quality management system (QMS) and/or relevant clause of AS9100 (see Figure B.8).

Nonconformity number	Location and process	Clause	Status (that is, major or minor)	Nonconformity
(*Note:* Each nonconformity must include the nonconformity, standard requirement, and objective evidence.)				

Figure B.8 Nonconformity chart.

- Identify opportunities for improvement without offering solutions.
- Use a format that includes root cause, corrective action, and systemic action. A sample Nonconformity Report can be seen in Figure B.9.

Opportunity for Improvement

Opportunities for improvement (OFI) don't require a formal response but may be revisited at future assessments.

STEP 13: CLOSING MEETING

The auditor is responsible for three things during step 13, the closing meeting:

- Determine audit team next steps.
- Prepare draft report.
- Conduct closing meeting.

Step 13a: Determine Audit Team Next Steps

Once all of the nonconformities are written, step 13a focuses on the audit team's recommendations. There are four outcomes:

- Minor and major nonconformities are so numerous that another readiness review and stage 2 on-site audit are required.
- Minor nonconformities that can be closed via written documentation.
- Minor nonconformities that require on-site closeout.
- Major nonconformities that require on-site closeout.

Step 13b: Prepare the Draft Report

- Prepare draft reports describing all nonconformities. Also, identify and include the audit team summary, at a minimum.
- Identify nonconformities and opportunities for improvement. No other categories are allowed.

Corrective Action Request			1-17-3 Revision B
Part A: Audit Information			
Department		Audit number	
Activity audited		CAR number	
Auditor		Date issued	
Auditee		Reference	
Part B: Nonconformity			
Nonconformity: Requirement: Objective evidence: Auditor Date Department representative Date			
Part C: Corrective/preventive action			
Immediate action Root cause Corrective action Auditor Date		Preventive action Department representative Date	
Part D: Verification of corrective action			
Follow-up details Auditor Date Department representative Date			

Figure B.9 Nonconformity report (NCR).

Step 13c: Conduct the Closing Meeting

A typical closing meeting agenda includes the following:

- Statement of thanks
- Attendance list
- Scope, objectives, and criteria
- Significance of audit samples
- Audit standard, rules, and reference manuals
- Audit summary
- Nonconformity statements, root cause, and systemic corrective action responses
- Opportunities for improvement
- Clarification of nonconformity statements and summary
- Statement of confidentiality
- Follow-up

STEP 14: AUDIT REPORT

Management review results and actions: ______________________________

____

____

Is the management review fully implemented and effective? A statement and summary is required for this question. If the answer is no, a detailed explanation is required.

Progress toward continual improvement: ______________________________

____

____

Internal audit results and actions: ______________________________

____

____

Customer complaints and organization resources: ____________________

____

____

Effectiveness of the management system in achieving customer and organizational objectives: ____

____

____

Review of new customers since last assessment (if applicable): _________

____

____

Structure of QMS and revision status: _____________________________

____

____

Include document changes since the last assessment.

About the Author

Chad Kymal is the chief technical officer and founder of Omnex Inc. Chad is an international trainer and consultant whose broad experience includes TQM, setup reduction, technology assessment and inventory analysis, statistical process control, and quality function deployment. After graduating from the General Motors Institute (GMI), Chad spent a number of years working at General Motors and KPMG before founding Omnex Inc. in 1986. Over Chad's successful career, he has served on the Malcolm Baldrige board of examiners and has received numerous quality achievement awards, including the Quality Professional of the Year award by the American Society of Quality Automotive Division in 2005. In addition to a bachelor's degree from GMI, Chad holds both a master's degree in industrial and operations engineering from the University of Michigan and an MBA from the University of Michigan.

In his role as chief technical officer, Chad is instrumental in determining the strategic direction of the consulting and training organizations, as well as the product strategy for the software company, Omnex Systems. Under Chad's direction, Omnex has worked with TRW, Ford, Philips Semiconductors, Magna, and most of the Fortune 500 companies worldwide. Chad personally led the Omnex team that helped Ford Motor Co. rewrite the QOS (quality operating system) methodology. Omnex provides more than 100 training courses in subject areas ranging from APQP to Six Sigma and has 15 offices worldwide.

Chad both developed and teaches Omnex's ISO 9001, ISO/TS 16949, AS9100, ISO 14001, OHSAS 18001, and integrated management systems (ISO 9001, ISO 14001, and OHSAS 18001) auditor courses, most of which are RABQSA certified. He has additionally led the development of Omnex's project management for new product/risk management (APQP), FMEA/control plan, SPC, PPAP (first article), and MSA courses.

Chad is a registered lead auditor who conducted the first worldwide witness audit for QS-9000. He helped develop QS-9000 and is one of the authors of the ISO/TS 16949 Semiconductor Supplement. He is also the founder of AQSR, one of the top five registrars as rated by *Quality Digest*. AQSR conducted certifications for ISO 9001; ISO/TS 16949; AS9100, AS9110, and AS9120; TL 9000; ISO 14001; OHSAS 18001; and many other international standards.

Chad has published numerous papers and three books on the subjects of management standards and quality. He is the author of four books on quality standards, including three on the auditing process. He is currently working on his next book, titled "Strategy for Risk Management in New Product Launch in the Aerospace Industry," which covers the implementation of risk management strategies in the aerospace industry.

Chad has conducted hundreds of audits in many different sectors—from aerospace to steel to distributors to electronics and semiconductors—in North America, Asia, and Europe over the last 20 years.

Chad has led the development of both lead and internal auditor courses that focus on both the aerospace industry and the AS9100/AS9101 standards. He is also leading two large implementations within the aerospace industry targeting product quality improvements for Bombardier Inc. (airplane) and Pratt & Whitney (engine). The Bombardier implementation involves working with every major system and subsystem supplier of the aircraft.

Index

A

B

C

D

E

N

O

P

Q

R

S

T

W

ASQ Membership

Research shows that people who join associations experience increased job satisfaction, earn more, and are generally happier*. ASQ membership can help you achieve this while providing the tools you need to be successful in your industry and to distinguish yourself from your competition. So why wouldn't you want to be a part of ASQ?

Networking

Have the opportunity to meet, communicate, and collaborate with your peers within the quality community through conferences and local ASQ section meetings, ASQ forums or divisions, ASQ Communities of Quality discussion boards, and more.

Professional Development

Access a wide variety of professional development tools such as books, training, and certifications at a discounted price. Also, ASQ certifications and the ASQ Career Center help enhance your quality knowledge and take your career to the next level.

Solutions

Find answers to all your quality problems, big and small, with ASQ's Knowledge Center, mentoring program, various e-newsletters, *Quality Progress* magazine, and industry-specific products.

Access to Information

Learn classic and current quality principles and theories in ASQ's Quality Information Center (QIC), *ASQ Weekly* e-newsletter, and product offerings.

Advocacy Programs

ASQ helps create a better community, government, and world through initiatives that include social responsibility, Washington advocacy, and Community Good Works.

Visit www.asq.org/membership for more information on ASQ membership.

*2008, The William E. Smith Institute for Association Research

ASQ Certification

ASQ certification is formal recognition by ASQ that an individual has demonstrated a proficiency within, and comprehension of, a specified body of knowledge at a point in time. Nearly 150,000 certifications have been issued. ASQ has members in more than 100 countries, in all industries, and in all cultures. ASQ certification is internationally accepted and recognized.

Benefits to the Individual

- New skills gained and proficiency upgraded
- Investment in your career
- Mark of technical excellence
- Assurance that you are current with emerging technologies
- Discriminator in the marketplace
- Certified professionals earn more than their uncertified counterparts
- Certification is endorsed by more than 125 companies

Benefits to the Organization

- Investment in the company's future
- Certified individuals can perfect and share new techniques in the workplace
- Certified staff are knowledgeable and able to assure product and service quality

Quality is a global concept. It spans borders, cultures, and languages. No matter what country your customers live in or what language they speak, they demand quality products and services. You and your organization also benefit from quality tools and practices. Acquire the knowledge to position yourself and your organization ahead of your competition.

Certifications Include

- Biomedical Auditor – CBA
- Calibration Technician – CCT
- HACCP Auditor – CHA
- Pharmaceutical GMP Professional – CPGP
- Quality Inspector – CQI
- Quality Auditor – CQA
- Quality Engineer – CQE
- Quality Improvement Associate – CQIA
- Quality Technician – CQT
- Quality Process Analyst – CQPA
- Reliability Engineer – CRE
- Six Sigma Black Belt – CSSBB
- Six Sigma Green Belt – CSSGB
- Software Quality Engineer – CSQE
- Manager of Quality/Organizational Excellence – CMQ/OE